# An Introduction to Fatigue
# in Metals and Composites

# An Introduction to Fatigue in Metals and Composites

R. L. Carlson
*Professor of Aerospace Engineering,*
*Georgia Institute of Technology, USA*

and

G. A. Kardomateas
*Associate Professor of Aerospace Engineering,*
*Georgia Institute of Technology, USA*

**CHAPMAN & HALL**
London · Glasgow · Weinheim · New York · Tokyo · Melbourne · Madras

**Published by Chapman & Hall, 2–6 Boundary Row, London SE1 8HN, UK**

Chapman & Hall, 2–6 Boundary Row, London SE1 8HN, UK

Blackie Academic & Professional, Wester Cleddens Road, Bishopbriggs, Glasgow G64 2NZ, UK

Chapman & Hall GmbH, Pappelallee 3, 69469 Weinheim, Germany

Chapman & Hall USA, 115 Fifth Avenue, New York, NY 10003, USA

Chapman & Hall Japan, ITP-Japan, Kyowa Building, 3F, 2-2-1 Hirakawacho, Chiyoda-ku, Tokyo 102, Japan

Chapman & Hall Australia, 102 Dodds Street, South Melbourne, Victoria 3205, Australia

Chapman & Hall India, R. Seshadri, 32 Second Main Road, CIT East, Madras 600 035, India

First edition 1996

© 1996 R.L. Carlson and G.A. Kardomateas

Printed and bound in Great Britain by
Hartnolls Limited, Bodmin, Cornwall

ISBN 0 412 57200 1

A catalogue record for this book is available from the British Library

Library of Congress Catalog Card Number: 95-071381

∞ Printed on permanent acid-free text paper, manufactured in accordance with ANSI/NISO Z39.48-1992 and ANSI/NISO Z39.48-1984 (Permanence of Paper).

# Contents

# Preface

The subject of fatigue has attracted the attention of engineers and scientists trained in a wide variety of disciplines. Metal physicists have studied the mechanisms associated with fatigue in single crystals of high purity metals. Metallurgists have investigated the effects of alloying, heat treatment and mechanical working on fatigue properties. Design engineers have focused attention on the development of strategies for predicting fatigue lifetimes. It is not too surprising, then, to learn that by 1983 at least 20000 papers had been published on the subject of fatigue (Pook, 1983). Early emphasis on fatigue studies was on metals. Interest in fatigue in polymeric materials, ceramics and composites has, however, increased steadily over the past 30 years. The literature on fatigue has, consequently, grown enormously as can be attested by the number of technical journals, conferences and symposia which are devoted to topics in fatigue.

The procedures used to design structural elements and machine components depend on the consequences of deviations from allowable performance. Sometimes, a restriction on deflection or displacement is imposed. For metals the limiting condition is often the onset of irreversible yielding at a critical location. In these examples the properties required are the elastic moduli and a yield condition. For metals these properties are generally available for the commercially produced metals which are often used. After the selection of a metal, design can proceed with applications of the theory of elasticity.

If other failure modes must be considered, a more comprehensive procedure is required. Such failure modes can include fracture and fatigue. These are dependent on component geometry, loading history, temperature and environment. An understanding of the resulting behavior can require a knowledge of not only the mechanics of problems, but also the microstructural features of the material to be used. This contention is supported by increasing instances of collaboration involving materials scientists and mechanics specialists.

The need for interdisciplinary approaches to the solution of design problems has been further dictated by an increasing interest in the use of composite materials. Composites are being fabricated in a wide variety of material combinations and forms. Composite materials with metallic, polymeric and ceramic matrices with fiber, filament and particle reinforcement are being produced. It is necessary, therefore, for design engineers to familiarize themselves not only with metals, but also with nonmetals and composites. Also, since for composites the details of the reinforcing elements can profoundly affect the mechanical properties, a knowledge of fabrication procedures is essential. Brief descriptions of fabrication processes are, therefore, included in our text. Since these descriptions cannot be comprehensive, however, references are included to provide readers with access to more detailed expositions.

Even by limiting emphasis to fatigue behavior, the scope of the topics addressed is still too great to be intensively covered. Many of the topics included could be the subjects of entire books. Thus, an effort has been made to present introductions to the topics which are included and to provide extensive reference listings for readers who wish to explore some of the subjects in greater depth.

The primary objective of this text is to present a balanced discussion of the phenomenon of fatigue in metals, nonmetals and composites. Design procedures currently in use are reviewed, along with discussions of the limitations of their application. The topic of structural integrity, as measured by use of damage tolerance methods, is examined and particular attention is devoted to discussions of the modes of fatigue failure which are characteristic of each of the materials considered.

One projected audience for the book includes engineers who are being required to consider the use of an expanding array of new materials. It has also been written for use as an introductory text both for students who wish to begin research investigations on fatigue subjects, and for those who are preparing for careers in the design of structures and machines. These latter objectives have been addressed by the inclusion of problems for each chapter at the end of the text. It is anticipated that for those courses for which a greater emphasis on some of the topics is desired, supplementary material from the listed references may be introduced. The list of problems which have been provided may then also be supplemented to provide an expanded coverage.

# Acknowledgements

The authors are indebted to many individuals who have had an influence on their careers. The first author enjoyed a continuing collaboration with the late Dr C. J. Beevers. The joint research which we undertook served as a stimulus for studies in fatigue crack growth behavior. The second author gratefully recognizes the benefits of being introduced to fracture research in mechanics by his doctoral program advisor, Prof. F. A. McClintock.

We must also acknowledge the permission granted by Drs M. D. Halliday and K. Tohgo to include photographs which effectively contribute to our discussions of fatigue crack growth in metals. We also are grateful to Dr T. Kevin O'Brien for giving us permission to use figures taken from his research on fatigue damage in polymeric matrix composite laminates.

The benefits of contributions from the three sons of the first author should be noted. Fracture in solids and the construction and use of composite materials were known about long before recorded history. The authors used sources on these topics provided by Prof. David L. Carlson. Currently, there is an increasing concern for the consequences of structural failures. We have provided a basis for consideration of legal aspects of this topic by the inclusion of an introductory discussion in Appendix A. Prof. Richard R. Carlson wrote this section. Lastly, Dr Robert L. Carlson, Jr provided us with access to some of the material which was used in the preparation of the chapter on biomaterials. Finally, our obligation to persons who made necessary contributions in the preparation of our text must be recognized. These include the wife of the first author, Betty Carlson, who typed the draft of our manuscript and our graduate student, David Hooke, who prepared figures. Julia Roach and Brian Henderson, assembled the final copy.

# 1

# Introduction

## 1.1 INITIAL RECOGNITION OF FRACTURE AND FATIGUE

People have, even in prehistoric times, demonstrated a remarkable ability to adapt available materials for useful applications. Archeologists have uncovered stone tools which indicate that they had begun to develop an empirical grasp of the mechanics of materials about 2 million years ago (Fagan, 1992). The tools and weapons which were subsequently developed over the period from 2 million to 10 thousand years ago manifest a utilitarian comprehension of the importance of such material properties as stiffness, strength and hardness (Fagan, 1992). Also, the development of hunting implements such as spears and the bow and arrow indicate that they were able to incorporate manifestations of the rudiments of dynamics in their designs. These examples demonstrate resourcefulness and an ability to improvise with available materials. This contention is further supported by their construction of shelters. In one example Paleolithic mammoth hunters of Russia arranged branches across the top of a pit, and then added layers of mud to the network to make an early form of a composite material (Hole and Heizer, 1965). Neither branches alone nor mud alone could have provided a satisfactory roof.

One of the ingredients in a philosophy of design is to anticipate modes of failure and to take measures to avoid them. One of the most widely recognized modes of failure is fracture. In an elegant paper entitled 'The Starting of a Crack', Eshelby (1969) begins by citing articles by Skertchley (1879) and Baden-Powell (1949) in which the intuitive grasp which 'flint-knappers' had for running cracks is described. The special fracture mechanisms of flint, which is a form of quartz, were recognized very early. Archeological research indicates that practical uses of crypto-crystalline rock such as flint may date back to at least 500000 years ago. Early humans discovered that they could exploit the fact that such rocks can be fractured in a controlled manner. They found that they could produce flakes which were useful for scraping operations, and shape knives, hand axes and arrow heads

(Fagan, 1992). The variety of mechanisms which were exploited to achieve the range of products which were made is impressive. In an interesting article Speth (1972) has analyzed techniques which could be used to create the types of products which have been found. He used results from the impact theory of wave propagation and concluded that spalling fractures could form the basis for the development of the variety of implements produced. A fascinating modern adaptation of the production of flakes has been described by Sheets (1993). He found that he could, using primitive techniques, produce obsidian (volcanic ash) flakes that had cutting edges which were much sharper than those of modern surgical scapels. Blades made from these flakes have been successfully used in delicate eye surgery. Because of the sharpness of the edges, more accurate incisions are possible, healing is faster and there is less scar tissue formation.

The useful forms of spalling which are characteristic of flint-like materials are not observed for all types of rock. The selection of flint, therefore, illustrates the importance of recognizing differences in fracture mechanisms.

The transition from the use of rock to metals initiated major improvements in both tools and weapons. Tylecote (1976) has written a comprehensive text on the history of metallurgy and he has summarized data on metal products which have been found by archeologists. The age of the products recovered varies considerably from one region to another, and only the earliest dates are given here. The earliest evidence of the use of copper dates to about 9000 years ago in Asia Minor. Implements and artifacts of bronze, an alloy of copper and tin, have been discovered in Iraq dating back 4800 years. The so-called Iron Age began about 4000 years ago in Asia Minor.

It was not until the scientific and mathematical advances of the 17th and 18th centuries that a foundation became available for the technological developments which occurred during the Industrial Revolution of the 18th and 19th centuries (Trevelyan, 1942; Seaman, 1982). This represented a transition period during which production techniques changed from the use of manual labor to the use of machines which were the products of engineering design.

Procedures which are used to design structural elements and machine components depend upon the consequences of deviations from allowable performance. In some cases a restriction on deflection or displacement is imposed. Often the limiting condition is the onset of irreversible yielding at a critical location. For these examples the material properties required are the elastic moduli and a yield condition. An analysis then proceeds with an application of the mechanics of deformable bodies.

If the imposed limiting condition is based on material fatigue or fracture, design requires a more comprehensive analytical procedure. Fatigue occurs under loads which vary continuously with time. It begins with the nucleation of a microcrack which extends during a growth period and finally terminates with an unstable, catastrophic fracture. Since machines and vehicles are commonly exposed to such loading histories, the phenomenon must be accounted for in design. Timoshenko (1953) suggests that Poncelet in 1839 may have been the first to use the term 'fatigue' to describe the fatigue process. An early use of the term fatigue to describe metal failures has been traced by Pook (1983) to Braithwaite (1854), who stated in a paper that 'many accidents in railways are to be ascribed to that progressive action which may be termed the fatigue of metals'. In response to failures which were diagnosed as being the result of fatigue many experimental research programs were conducted in the latter part of the 19th century and into the 20th century. Pook (1983) has estimated that by 1983 at least 20000 papers had been published on the subject of fatigue. A good summary of progress in fatigue is presented in a text by Fuchs and Stephens (1980).

The fracture of inherently brittle materials, such as glass, has been relatively well understood since the classic research of Griffith (1920). The development of an understanding of brittle fractures of nominally ductile metals, however, did not emerge until the work of Irwin (1948) and Orowan (1952). Furthermore, the identification of the basic driving force for fatigue crack growth was not discovered until conjectures offered by Paris, Gomez and Anderson (1961) were generally accepted. Thus, although people were able to recognize and exploit fracture phenomena at least 2 million years ago, they only have relatively recently begun to understand the mechanisms which govern fracture and fatigue. Also, from the extensive research which is currently being conducted, it is clear that some fundamental issues remain to be resolved.

From a review of the history of fatigue and fracture, it can be concluded that the progress to date has been driven by attempts to provide answers to two legitimate but distinct questions. These are:

1. How can the responses observed be described, not only qualitatively, but quantitatively?

2.     What are the explanations for the responses observed for the
       variety of loading conditions that can be encountered?

Emphasis in much of early fatigue research, and even ongoing efforts
in the development of design codes, may be described as efforts to find
answers to the first question. The second question focuses attention on
the identification of operative mechanisms. In the chapters which follow
both the features of fatigue behavior which are reasonably well
understood, and issues which remain to be resolved are discussed.

## 1.2 EVOLUTION OF TESTING PROCEDURES

Although the occurrence of some forms of fatigue were probably not
uncommon during the bronze and iron ages, it was not until the middle
of the 19th century that fatigue was recognized as a problem which had
to be addressed by engineers. The primary sources of concern were
incidents in which fatigue failures were occurring in railway axles. The
increased use of rail transportation led, particularly in Germany and
England, to a general consensus that experimental studies which could
provide data for design limitations should be initiated. It is of interest to
note that a very early concern for legal liability surfaced. In a meeting of
the British Institution of Mechanical Engineers the chairman, Robert
Stephenson (1849), concluded a discussion of fatigue failures by saying
'I am only desirous to put the members on their guard against being
satisfied with less than incontestable evidence as to a molecular change
in iron, for the subject is one of serious importance, and the breaking of
an axle has on one occasion rendered it questionable whether or not the
engineer and superintendent would have a verdict of manslaughter
returned against them'.   Thus, the concern for legal liability that
currently attends failures in which injuries or loss of life occur has a
history which goes back many years. Product liability, which was an
isolated matter of concern for Stephenson in 1849, has in more recent
times become recognized as an issue of widespread importance. The
evolution of product liability law and the current status in Europe and
the United States are briefly reviewed in Appendix A.
   Early experimental investigators of fatigue behavior very logically
attempted to duplicate as nearly as practically possible the conditions
which were developed in the service incidents which led to failures.
Wöhler in Germany, for example, initially conducted fatigue tests on full

size railway axles. Subsequently, he conducted tests on small specimens, and designed machines which produced cyclic bending, reversed bending, uniaxial loading and torsion. A concise review of Wöhler's contributions appeared in an article in the August 23, 1867 issue of the British weekly journal, *Engineering* (pp. 160 – 1). Descriptions of other early work are given in texts by Timoshenko (1953) and Moore and Kommers (1927). A more recent summary which also contains descriptions of both features of modern testing systems and commonly used test specimens is contained in a text by Fuchs and Stephens (1980).

Fatigue tests are conducted on full scale components such as an aircraft wing attached to a fuselage in order to duplicate the complex diffusion of stress into critical areas. They are sometimes conducted on sub-assemblies and components which are parts of a total structural system such as a vehicle, a bridge or a machine on a production assembly line. Finally, they are often conducted on small specimens which are designed to provide fatigue data which reveal the effects of stress state, surface preparation, environment and loading history on a material of interest.

Prior to about 1960 the primary goal of fatigue testing was to obtain stress versus cycles to failure data on specimens which initially were nominally free of cracks. There has been a widespread increase in the use of precracked specimens since then, however, and the data obtained have been used in conjunction with an application of fracture mechanics to form the basis for the development of a new philosophy of design. The goal of this philosophy is to determine the load carrying 'tolerance' of a component in which a crack is present.

The growing interest in fatigue crack propagation in the 1960s coincided with a rapid spread in the use of servo-hydraulic testing systems. The versatility of these systems cannot be matched by any of the testing machines which were previously available. For these new systems a hydraulic jack or actuator is controlled by a servo-valve which responds to the differential feedback from a sensor in a closed loop system. The sensor can be a load cell or a displacement transducer. Thus, either load or displacement control is possible. Also, in addition to constant amplitude testing, variable amplitude testing can be programmed through a controlling digital computer. Uniaxial tests can be conducted in commercially available load frames, and fixtures can be used to perform bending and torsion tests. Loading frames for biaxial tests are also available.

The use of the servo-hydraulic system is not restricted to use in loading frames. Hydraulic actuators can also be used in specially designed testing frames which apply cyclic loading to sub-assemblies or to such a complex structure as a complete aircraft. Coordinated, multiple actuators are commonly used in such applications. Descriptions of a variety of uses of servo-hydraulic testing systems are given in a review article by Marsh and Smith (1986).

The emphasis on fatigue crack tests has increased interest in non-destructive testing techniques for detecting cracks. Periodic inspections for detecting the presence of cracks in structural systems during service have become common practice for such vehicles as aircraft. Techniques for measuring the growth of cracks are also a crucial part of testing procedures in laboratory fatigue crack tests.

## 1.3  SCOPE OF FATIGUE IN SOLIDS

It may be inferred from the preceding discussion that the term fatigue, as applied to the consequences of cyclic loading, is exclusively associated with a response observed in metals. Although the mechanisms differ, fatigue damage has also been found to occur in other solids. To provide a perspective for the scope of fatigue phenomena which have been identified, a brief summary of the qualitative features of fatigue in both metals and nonmetals serves as an introduction to the more detailed presentations which are provided in subsequent chapters.

The fatigue of metals has long been characterized by the use of plots of stress range versus the log of cycles to failure as depicted in Fig. 1.1. In this representation the minimum stress on a specimen is zero so that the stress range is then equal to the maximum stress. Although the nucleation of the crack which ultimately results in failure cannot be readily discerned, it is possible to represent it, at least qualitatively, by the lower curve. Let the upper curve then represent total failure. The horizontal line A – B then represents all phases of fatigue life for the given maximum stress, i.e. crack nucleation, crack growth and final, catastrophic failure. Also, if a line from point A never intersects the upper curve for a given maximum stress, the initiated crack can be described as non-propagating or arrested. The issue of whether or not the upper curve has a horizontal asymptote is discussed in Chapter 4. If a fatigue crack does not arrest after initiation, it can grow incrementally under continued load cycling. Until abrupt failure occurs, however, the continuing extension is described as 'stable' fatigue crack growth. A visual examination of the fatigue portion of a fractured surface indicates that stable growth is nominally brittle, i.e. there is no evidence of gross plastic flow. The operative mechanisms which produce the features exhibited by the fatigue portion of the fractured surface depend upon the rate of crack growth (section 3.2). When the applied loading conditions and the crack length become critical, an abrupt fracture occurs during a final loading cycle and the portion of the cross-section remaining is more characteristic of that which is observed for monotonic loading to failure.

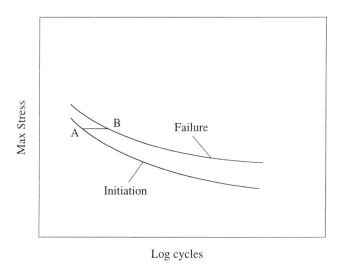

**Fig. 1.1** Fatigue crack initiation and failure curves

The appearance of the fatigue crack surface of a failed component makes it possible to determine the location of the site at which initiation occurred. Partial, ring-like markings, which are called 'beach marks' are centered about the initiation site. Often this site is located at a small flaw which was introduced during manufacturing or servicing.

The appearance of beach marks is illustrated in Fig. 1.2. In each of the four figures the last growth ring represents the onset of critical unstable crack extension. Figs. 1.2 (a) and (b) are corner and surface cracks in bars with a rectangular cross-section. The cross-section in Fig. 1.2 (c) illustrates a case in which fatigue cracks have started at two sites, i.e. at diagonal corners. Fig. 1.2 (d) represents a case in which a surface crack was initiated on the periphery of a circular, cylindrical bar. The beach marks separate regions where the growth has been arrested and restarted or where the growth rates or the corrosion conditions differ. The elements of fracture mechanics can be used to analyze fatigue fracture surfaces and reconstruct the events which have led to failure (Hertzberg 1989). A detailed discussion of the appearance of fatigue crack surfaces has been presented by Peterson (1950).

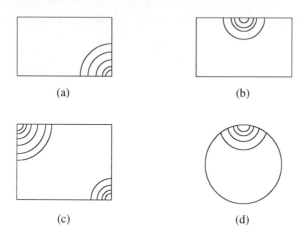

**Fig. 1.2** Beach marks for different types of fatigue cracks

The conditions encountered in service are often more complex than those depicted in Fig. 1.1. The maximum stress and minimum stress, for example, often vary from cycle to cycle in what is described as variable amplitude loading. Also, corrosion effects are sometimes present. If a crack is propped open and a specimen is exposed to a corrosive medium, the crack can grow. If cyclic loading is applied, a coupled behavior described as corrosion fatigue ensues.

If a loaded cracked component is exposed to an absolute temperature that is greater than about one-half the absolute melting temperature, creep crack growth can occur. If the loading varies with time, as in fatigue, a coupled crack growth behavior involving both creep and fatigue can occur.

Research investigations which have contributed to an understanding of these service connected problems have been conducted and efforts have been made to develop methods which can be used to predict the resulting behavior. Descriptions of the features of these problems and discussions of issues which remain to be resolved are presented in subsequent chapters.

polymeric solids is very distinct from that of metals, they qualitatively exhibit similar responses to some external stimuli. Stress versus cycles to failure curves similar to those for metals are generated and a stress below which failure does not occur may be observed. Fatigue crack growth tests have also been conducted on engineering plastics, and correlations of the type used for crack growth in metals have been applied.

The fracture of brittle solids has long been of interest because of the widespread manufacture and use of ceramic articles. Investigations of fatigue in inherently brittle solids such as glass and ceramics have also been conducted. McClintock and Argon (1966) have summarized a behavior observed in glass which has been described as 'static' fatigue. Prolonged loading has been found to result in fractures at strengths smaller than those obtained for rapid loading. The behavior has been attributed to a corrosive attack by water vapor, and does not require cyclic loading. It is not, therefore, appropriate to describe this as a fatigue phenomenon.

Another class of solids which is of widespread interest involves mixtures of materials. The term composites has been used to represent these multiphase engineering solids. Fiber-reinforced composites with plastic, metal and ceramic matrices have been fabricated and some mechanical properties have been evaluated. The variety of fibers, matrices, and the combinations of fiber orientation and fiber volume fractions which are possible is immense, however, so it is difficult to provide characterizations which are generally valid. Although fatigue damage has been observed, and failure mechanisms have been identified, it has been found that the behavior is dependent not only upon the fiber – matrix combinations, but also upon the type of loading applied. Since these details can vary substantially, it is expedient to defer further discussion of the fatigue properties of composites to subsequent sections in which specific examples will be examined.

## 1.4 DEVELOPMENT OF DESIGN METHODS

The conception of new designs requires not only an analysis of structural configurations and loading conditions, but also a consideration of possible modes of failure. Fairburn (1864) in a discussion of bridge designs observed that the initial load capacity of tubular bridges had been selected to 'be six times the heaviest load that could ever be laid upon them after deducting half the weight of the tube. This was

upon them after deducting half the weight of the tube. This was considered a fair margin of strength, but subsequent considerations such as generally attend a new principle of construction with an untried material induces an increase of strength and, instead of the ultimate strength being six times, it was increased to eight times the weight of the greatest load.' Although this may appear to have been a grossly conservative prescription for design, it actually was a rational decision considering the uncertainties associated with the structural analysis, the loading conditions and the material properties. In a properly conceived design the 'margin of safety' should in fact provide confidence in the integrity of the structure, vehicle or machine during its service lifetime.

The design process has, since Fairburn's comments, become more sophisticated, but the principal objectives have remained essentially unchanged. When anticipating fatigue as a failure mode, it has been common practice for many years to require that a structure should be able to survive several times the intended service lifetime, e.g. the time to failure of an aircraft wing in a laboratory fatigue test may be required to be four times the expected service lifetime. This type of requirement is a basic feature of a design philosophy described as **safe life**.

There are designs in which the failure of a single component results in a total system failure. When possible, however, it is obviously desirable to create a design in which there is a shifting of the function of a failed component to other components. As a simple example, consider a two bar truss under an external load. If one bar fails, the remaining unfailed bar can no longer serve the intended structural function. By contrast the failure of one bar in a three bar truss simply results in a redistribution of the bar forces and the truss can still, at least temporarily, serve its intended function. The latter example can then be described as being **fail safe.** This is obviously a desirable design solution and it is commonly used in aircraft.

The presence of a crack in a component subject to load variations does not necessarily constitute failure. The crack may undergo a time-dependent extension which is often described as 'stable' growth. Eventually, the growing crack may attain what is described as a critical length and then unstable or catastrophic growth can occur. Thus, during a stable growth period, the structural integrity of the system remains intact and the primary concern is the anticipation of when a critical length will be attained. This requires a knowledge of the loading history on the cracked component. Also, to forestall the onset of unstable growth, the crack should be monitored at prescribed inspection intervals.

Although the task described is complex and a simplified description has been presented, the logic involved focuses attention on an attempt to evaluate the tolerance of components to the presence of cracks. The elements of the procedure outlined have formed a basis for the evolution of a sophisticated procedure which is described **damage tolerance design**.

In view of the many uncertainties associated with fatigue behavior the task of managing a fleet of aircraft by anticipating how and when to perform inspections and make repairs or modifications may seem overwhelming. In response to the need to be able to control such operations, however, there has been a growing interest in the evolution of strategies which can provide a basis for the development of fleet management capabilities. This has required the identification of the primary statistical parameters which are entailed. It has also focused attention on the need to examine issues of fatigue behavior which remain to be resolved. More detailed discussions of these topics are presented in subsequent chapters.

# 2

# Elements of deformable body analysis

## 2.1 INTRODUCTION

Forces applied to solids which are defined as being rigid do not cause any changes in shape or size. Solids used as structural elements are not rigid, however, and they are deformed when loaded. The subjects of elastic, inelastic and time-dependent deformable solids are comprehensively developed in a number of texts. Only the elements which are required for an introduction to the topics which are included in subsequent chapters are addressed here. Where results from specialized treatments are used, appropriate references are provided.

## 2.2 LOADING MODES

Monitoring the response of a deformable body to externally applied loads would seem to be a relatively simple task. Upon closer examination, however, it is found that alternative options for applying loads must be recognized. Also, a consideration of data from a variety of experiments reveals that the features of the responses for different loading options are not always unique. To establish a sound basis for reviewing experimental results which are of interest here, a brief description of loading options and the types of responses that can be observed for alternative choices is presented.

A simple means of conducting a test to determine a response behavior is to apply either a force or a displacement which is continuously increased. As long as certain conditions are satisfied, the behavior observed for the two options will be the same. To emphasize the importance of distinguishing between the two methods of loading, however, consider a response for which the behavior differs in spite of the fact that the reaction is elastic and reversible. This is illustrated in the so-called snap-through buckling problem which is typified by the action which occurs when the bottom of an oil can is pushed inward. The alternative responses are illustrated in Fig. 2.1.

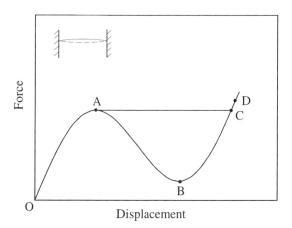

**Fig. 2.1** Snap-through buckling

The response for either force or displacement control is the same from point O to point A. Beyond point A, however, the response snaps through along the line segment A – C and then moves upward along curve C – D for force control. The force beyond point A can, however, decrease for deplacement control as shown by the curve A – B – C – D. Points on the diagram represent load equilibrium states. Upon removal of load, the equilibrium state is represented by point O again.

The instability behavior described is not restricted to elastic buckling phenomena. Bars subjected to tensile loading beyond the elastic limit can exhibit a relative maximum load beyond which the response behavior differs depending on the type of loading applied. The upper and lower yield points observed for low carbon steels lead to response features which are qualitatively similar to those illustrated in Fig. 2.1. Also. the maximum load or 'ultimate load' which develops in many metals is another example in which a relative maximum can give rise to behavior which is loading mode dependent. It may be concluded, therefore, that the type of behavioral response developed can depend not only upon the test material, but also upon characteristics of the loading process.

In essence the examples cited possess features which are common to many stability problems. It will become clear, moreover, that fracture and fatigue crack growth both exhibit features which can appropriately be described as involving transitions from stable to unstable behavior. The type of loading which is chosen for a given application depends upon the objectives of the experiments which are being conducted. Often stress and strain, rather than force and displacement, are used to display test data. It is, therefore, expedient at this point to introduce the definitions which are used here for stress and strain.

Stress provides a measure of the intensity of the loading state produced by applied loads. For the unidirectional loading of a cylindrical bar the stress, $\sigma$, may be defined by the equation

$$\sigma = \frac{F}{A},$$

(2.1)

where $F$ is the applied force and $A$ is the cross-sectional area normal to the load. The area $A$ changes as the force is applied, and it is given by the equation

$$A = A_0 + \Delta A, \quad \Delta A = A_0 - A$$

(2.2)

where $A_0$ is the unloaded, cross-sectional area, and $\Delta A$ is the change due to the applied force. Equation (2.2) may also be written as

$$A = A_0 (1 + \frac{\Delta A}{A_0}). \quad A = A_0 - \Delta A$$

(2.3)

It follows that if

$$\Delta A / A_0 \ll 1,$$

(2.4)

$A$ in equation (2.1) may be set equal to $A_0$.

The total strain, $\varepsilon$, may be defined as

$$\varepsilon = \int_0^\varepsilon d\varepsilon = \int_{L_0}^L \frac{dL}{L} = \ln \frac{L}{L_0},$$

(2.5)

where $L$ is the current, loaded length and $L_0$ is the initial, unloaded length. By expanding equation (2.5) in a power series expansion, it can be rewritten as

$$\varepsilon = \frac{\Delta L}{L_0}\left[1 - \frac{1}{2}\left(\frac{\Delta L}{L_0}\right) + \frac{1}{3}\left(\frac{\Delta L}{L_0}\right)^2 - \cdots\right].\tag{2.6}$$

For

$$1 \gg \left|\frac{1}{2}\left(\frac{\Delta L}{L_0}\right) - \frac{1}{3}\left(\frac{\Delta L}{L_0}\right)^2 + \cdots\right|,\tag{2.7}$$

the strain may be defined as

$$\varepsilon = \frac{\Delta L}{L_0},\tag{2.8}$$

where $\Delta L = (L - L_0)$. The definitions obtained by use of equations (2.4) and (2.7) are often described as engineering stress and engineering strain, respectively. Values based on equations (2.1) and (2.5) are described as true stress and natural or logrithmic strain, respectively (Johnson and Mellor, 1972). These measures of stress and strain are of interest for low cycle fatigue applications and in the determination of mechanical properties for rolling, drawing, extrusion and forging operations.

The most frequently used mode of loading is that for which the elastic and inelastic properties are being sought. The properties required for metals are the elastic moduli, the onset of inelastic behavior and the rate of strain hardening. These properties are depicted in Fig. 2.2, where the linear, elastic modulus

$$E = \frac{d\sigma}{d\varepsilon},\tag{2.9}$$

and the rate of strain hardening beyond the linear, elastic limit is given by the tangent modulus

$$E_{\mathrm{T}} = \frac{d\sigma}{d\varepsilon}.\tag{2.10}$$

For brittle materials fracture often occurs within the elastic range.

For polymeric materials and metals at elevated temperatures the stress – strain curves can become strain rate dependent. Time dependent strain or creep is then observed and the values of stress obtained for a given strain decrease with decreasing strain rate.

Stress – strain curves for composite materials have features which depend on the constituents and the fabrication details. These issues are discussed in Chapter 3.

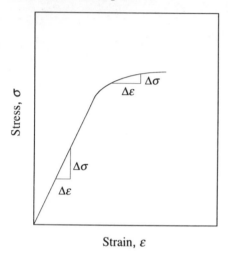

Strain, $\varepsilon$

**Fig. 2.2**  Engineering stress versus engineering strain

The simplest modes of loading used in fatigue tests involve variations of either stress or strain. The stress values which describe the cyclic stress history are indicated in Fig. 2.3. Two loading parameters which are used are, by definition, the stress range

$$\Delta\sigma = \sigma_{max} - \sigma_{min},$$
(2.11)

and the stress ratio

$$R = \frac{\sigma_{min}}{\sigma_{max}}.$$
(2.12)

Since $\sigma_{min}$ can be chosen to be compressive, it follows that $R$ can be negative.

The quantities used to describe strain cycling are shown in Fig. 2.4. The diagram represents tests which are conducted under constant amplitude, fully reversed cycles of strain. Since the loading level applied exceeds the elastic limit, the strain range, $\Delta\varepsilon$, is composed of both elastic and inelastic components.

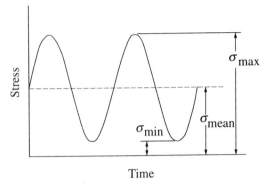

**Fig. 2.3**  Cyclic variation of stress

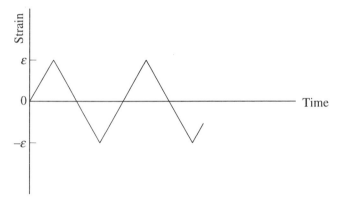

**Fig. 2.4**  Cyclic variation of strain

In some types of machinery the cyclic loading encountered is reasonably well represented by the simple loading program depicted in Fig. 2.3. Often, however, variable amplitude loading is developed. Extreme examples of this departure from constant amplitude loading occur in aircraft. In these examples the type of aircraft and the kind of mission being flown result in characteristic load spectra. These have been monitored to provide load data which can be used in testing programs.

Examples of spectra which have been developed for use in the laboratory include (ten Have, 1989):

fighter aircraft lower wing skins – FALSTAFF
transport aircraft lower wing skins – TWIST
helicopter rotor blades – HELIX, FELIX
tactical aircraft cold end engine disks – Cold TURBISTAN
tactical aircraft wing skin composites – ENSTAFF
tactical aircraft hot end engine disks – Hot TURBISTAN
horizontal axis wind turbine blades – WISPER
off shore structures – WASH

Issues which arise with variable amplitude loading are discussed in Chapters 4 and 5.

The conditions encountered in service applications often include environments which introduce behavior that can interact deleteriously with the mechanisms commonly associated with fatigue. One example involves active atmospheres which promote corrosion. Although exposure to salt and acid solutions are readily accepted as being detrimental, it has also been found that normal atmospheres can have an influence on fatigue life. This has been demonstrated by comparisons of results from tests conducted in air and in a vacuum.

A second example of environment involves temperature. When the absolute operating temperature of a metal is greater than about one-half of the absolute melting temperature, time dependent deformation or creep can occur. This condition, which is encountered in gas turbine engines, for example, can lead to creep – fatigue interaction. Frequency and dwell time effects can then be exhibited. Corrosion – fatigue and creep – fatigue interactions are discussed in Chapters 4 and 5.

## 2.3 GOVERNING EQUATIONS

### 2.3.1 Introduction

Solutions to problems in the mechanics of deformable bodies require consideration of three ingredients which form the basis for the governing equations. These are:

1. Each infinitesimal element within the body must be in a state of equilibrium .

2. A relationship between the kinematics of displacement and a measure of strain must be specified at every point.
3. A constitutive law which relates stress and strain or strain rate must be specified.

If the behavior of interest involves large strains, all of the equations derived from the above ingredients are nonlinear. The behavior of interest in this text is primarily concerned with small strains, and the first two ingredients lead to equations which are linear. For small strains and linear elastic behavior all three sets of governing equations are linear. Nonlinear equations which characterize large strain behavior are presented by Sokolnikoff (1956) and Green and Zerna (1968).

## 2.3.2 Equilibrium equations

The requirements for equilibrium at a point in a body can be visualized by reference to the small element, $\Delta x_1 \, \Delta x_2 \, \Delta x_3$, shown in Fig. 2.5. The faces of the element are normal to the orthogonal coordinate axes $x_1, x_2$ and $x_3$. The body has been cut on interior planes which are normal to the axes. The directions of the stresses acting on the forward face are indicated. Note that the first subscript indicates the direction of the normal to this face. The second subscript gives the direction of the indicated stresses.

In addition to $\sigma_{11}$, two other stresses produce forces in the $x_1$ direction. These are the shear stresses $\sigma_{21}$ and $\sigma_{31}$.

In the absence of body forces, summing the forces produced by these stresses in the $x_1$ direction and letting $\Delta x_1$, $\Delta x_2$ and $\Delta x_3$ pass to zero gives the partial differential equation

$$\frac{\partial \sigma_{11}}{\partial x_1} + \frac{\partial \sigma_{21}}{\partial x_2} + \frac{\partial \sigma_{31}}{\partial x_3} = 0. \tag{2.13}$$

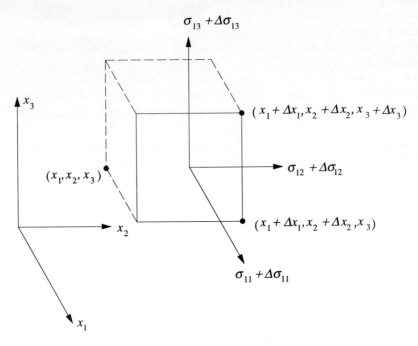

**Fig. 2.5**   Stresses on an element within a loaded body

Summing forces in the $x_2$ and $x_3$ directions gives

$$\frac{\partial \sigma_{12}}{\partial x_1} + \frac{\partial \sigma_{22}}{\partial x_2} + \frac{\partial \sigma_{23}}{\partial x_3} = 0 \qquad (2.14)$$

and

$$\frac{\partial \sigma_{13}}{\partial x_1} + \frac{\partial \sigma_{23}}{\partial x_2} + \frac{\partial \sigma_{33}}{\partial x_3} = 0. \qquad (2.15)$$

By summing moments about the three axes directions, it follows that

$$\begin{aligned} \sigma_{21} &= \sigma_{12} \\ \sigma_{13} &= \sigma_{31} \\ \sigma_{32} &= \sigma_{23}. \end{aligned} \qquad (2.16)$$

It can be inferred from equations (2.16) that equations (2.13), (2.14) and (2.15) provide the required conditions for equilibrium in terms of six stress variables.

If couple stresses act on the element faces, equations (2.16) are modified. Couple stresses may be visualized as being the result of distributed moments and torques per unit area which are applied to elemental areas such as $\Delta x_2 \, \Delta x_3$. If couple stresses are present, equations (2.16) are no longer valid. The alteration in these equations can be revealed by solving problem 2.9. The general consequences of this complication are discussed by McClintock and Argon (1966).

### 2.3.3 Strain – displacement equations

The dependent variables in the mechanics of deformable bodies are stress, displacement and strain. The coupling of these variables requires that the kinematics of deformation be expressed in relations between strain and displacement. A variety of measures of strain have been proposed. A valid measure of strain which describes large strains should, however, provide acceptable linear relations when strains become small. A measure of strain which satisfies this requirement is expressed as

$$ds^2 - ds_0^2 , \tag{2.17}$$

where the unloaded and loaded elemental lengths are $ds_0$ and $ds$, respectively. Changes in both elemental length and rotation are incorporated in this measure. It can be shown (Fung, 1965) that

$$ds^2 - ds_0^2 = 2\varepsilon_{ij} dx_i dx_j, \tag{2.18}$$

where $\varepsilon_{ij}$ are components of the Almansi strain tensor. The $x_i$ and $x_j$ are coordinates in the deformed state.

The strain $\varepsilon_{11}$ in terms of displacements $u_1$, $u_2$, and $u_3$ in rectangular Cartesian coordinates is

$$\varepsilon_{11} = \frac{\partial u_1}{\partial x_1} - \frac{1}{2}\left[\left(\frac{\partial u_1}{\partial x_1}\right)^2 + \left(\frac{\partial u_2}{\partial x_1}\right)^2 + \left(\frac{\partial u_3}{\partial x_1}\right)^2\right], \tag{2.19}$$

with similar equations for $\varepsilon_{22}$ and $\varepsilon_{33}$.

Also,

$$\varepsilon_{12} = \frac{1}{2}\left[\frac{\partial u_1}{\partial x_2} + \frac{\partial u_2}{\partial x_1} - \left(\frac{\partial u_1}{\partial x_1}\frac{\partial u_1}{\partial x_2} + \frac{\partial u_2}{\partial x_1}\frac{\partial u_2}{\partial x_2} + \frac{\partial u_3}{\partial x_1}\frac{\partial u_3}{\partial x_2}\right)\right],$$  (2.20)

with similar equations for $\varepsilon_{13}$ and $\varepsilon_{23}$.

   When the squares and products of the derivatives are negligible, the above definitions reduce to the Cauchy infinitesimal strain definitions, and

$$\varepsilon_{11} = \frac{\partial u_1}{\partial x_1} \ , \ \ \varepsilon_{22} = \frac{\partial u_2}{\partial x_2} \ , \ \ \varepsilon_{33} = \frac{\partial u_3}{\partial x_3}$$  (2.21)

and

$$\varepsilon_{12} = \varepsilon_{21} = \frac{1}{2}\left(\frac{\partial u_1}{\partial x_2} + \frac{\partial u_2}{\partial x_1}\right),$$

$$\varepsilon_{13} = \varepsilon_{31} = \frac{1}{2}\left(\frac{\partial u_1}{\partial x_3} + \frac{\partial u_3}{\partial x_1}\right),$$  (2.22)

$$\varepsilon_{23} = \varepsilon_{32} = \frac{1}{2}\left(\frac{\partial u_2}{\partial x_3} + \frac{\partial u_3}{\partial x_2}\right).$$

Note that the tensor components of equation (2.22) are related to the shear strains, $\gamma_{ij}$, as follows:

$$\varepsilon_{12} = \frac{1}{2}\gamma_{12},$$

$$\varepsilon_{13} = \frac{1}{2}\gamma_{13}, \qquad (2.23)$$

$$\varepsilon_{23} = \frac{1}{2}\gamma_{23}.$$

The displacements can be determined by integration of equations (2.21) and (2.22). By requiring that displacements be single valued for integration between two points along different paths, six compatibility equations can be derived.

The satisfaction of the compatibility equations is a necessary and a sufficient condition for single valued displacements in a simply connected body (no holes). For a multiply connected body it is a necessary but no longer sufficient condition for single valued displacements and additional integral relations must be satisfied (Fung, 1965).

For most of the problems of interest here the solutions can be formulated in terms of two dimensional elasticity. The compatibility equation which is then applicable is

$$\frac{\partial^2 \varepsilon_{11}}{\partial x_2^2} + \frac{\partial^2 \varepsilon_{22}}{\partial x_1^2} - 2\frac{\partial^2 \varepsilon_{12}}{\partial x_1 \partial x_2} = 0. \qquad (2.24)$$

### 2.3.4 Linear elastic constitutive laws

A homogeneous linear elastic body is one for which the strain, during loading, is linearly proportional to the stress. An isotropic elastic body has elastic properties which are the same in all directions. If the elastic properties differ in different directions, the body is described as being anisotropic. Single crystals and wood are examples of solids which are anisotropic.

Two types of solids are of interest here. These are the isotropic and the orthotropic solids. The orthotropic solid is a special case of general anisotropy, and its properties are the same for a 180° rotation about any one of three orthogonal coordinate axes.

Metals are often assumed to be both homogeneous and isotropic. The validity of the assumed isotropy depends on the size of the grains and the randomness of their crystallographic orientations. If the scale of observation involves a volume which is of the order of the grain size, isotropy can no longer be assumed. Also, various fabrication processes such as drawing and rolling can introduce preferred orientation directions of grains and hence invalidate the assumption of isotropy. Examples of orthotropy include wood, rolled metal, and lamina in fiber reinforced composites.

If coordinate axes are chosen parallel to the axes of structural symmetry in an orthotropic solid, the constitutive relations are given by Hooke's law, which can be written as

$$
\begin{aligned}
\varepsilon_{11} &= S_{11}\sigma_{11} + S_{12}\sigma_{22} + S_{13}\sigma_{33} , & \gamma_{12} &= S_{44}\sigma_{12}. \\
\varepsilon_{22} &= S_{21}\sigma_{11} + S_{22}\sigma_{22} + S_{23}\sigma_{33} , & \gamma_{23} &= S_{55}\sigma_{23}. \\
\varepsilon_{33} &= S_{31}\sigma_{11} + S_{32}\sigma_{22} + S_{33}\sigma_{33} , & \gamma_{31} &= S_{66}\sigma_{31}.
\end{aligned}
\tag{2.25}
$$

The elastic constants with mixed, interchanged subscripts are equal, i.e.,

$$ S_{12} = S_{21},\ S_{13} = S_{31},\ S_{23} = S_{32}. $$

Note that there is no interaction between either the shear components or the shear and normal components for this orientation of the coordinate axes.

For an isotropic, linear elastic solid Hooke's law can be written in the form

$$
\begin{aligned}
\varepsilon_{11} &= \tfrac{1}{E}\left[\sigma_{11} - v(\sigma_{22} + \sigma_{33})\right] \\
\varepsilon_{22} &= \tfrac{1}{E}\left[\sigma_{22} - v(\sigma_{33} + \sigma_{11})\right] \\
\varepsilon_{33} &= \tfrac{1}{E}\left[\sigma_{33} - v(\sigma_{11} + \sigma_{22})\right] \\
\gamma_{12} &= \frac{\sigma_{12}}{\mu} \\
\gamma_{13} &= \frac{\sigma_{13}}{\mu} \\
\gamma_{23} &= \frac{\sigma_{23}}{\mu},
\end{aligned}
\tag{2.26}
$$

where the coefficients of equation (2.25) can be shown to be

$$S_{11} = S_{22} = S_{33} = 1/E$$

$$S_{12} = S_{13} = S_{23} = -v/E$$

$$S_{44} = S_{55} = S_{66} = 1/\mu$$

and $E$ is Young's modulus, $v$ is Poisson's ratio and $\mu$ is the shear modulus. Also, for an isotropic solid the three elastic constants are interrelated by the relation

$$\mu = \frac{E}{2(1+v)}. \tag{2.27}$$

The elastic properties of fiber – matrix composites depend upon the geometry, the distribution and the volume fraction of the fiber – matrix mixture, and on the elastic properties of the two components. A good introduction to methods for calculating the elastic properties for composites with unidirectional and randomly oriented continuous fibers, and with short fibers is presented in chapter 5 of Hull (1981).

Laminate composites are not homogeneous, and the individual lamina in a layup often have different orientations. Analyses of composite laminates require that the elastic properties be derived by coupling the individual laminae together. The elastic properties of each lamina, the relative orientations and the compatibility of the interlaminar interfaces are a part of this analysis. The elements of the required analyses are presented in chapter 6 of Hull (1981). Additional sources may be found in texts by Jones (1975) and by Ashton, Halpin and Petit (1969).

## 2.4 THE CRACKED BODY PROBLEM

### 2.4.1 Introduction

Early in the twentieth century elasticians began solving problems involving voids or holes in loaded bodies. The cases studied included circular and elliptical holes in plates and ellipsoidal holes within solids (Timoshenko, 1953). Griffith (1920), who was interested in the effect of cracks on fracture, adopted results from analyses performed by Inglis (1913) to analyze test data which he obtained on the fracture of glass tubes and spheres. Inglis had solved the problem of a tensioned sheet with an elliptical hole, and Griffith considered the special case in which the ratio of the minor axis to the major axis passed to zero. The results which Griffith obtained were of special interest in investigations of the fracture of brittle materials such as glass and ceramics.

It was not until the early 1940s that the concepts introduced by Griffith began to be considered for application to the fracture of metals. The interest in the extension to metals stemmed from the occurrences of a number of catastrophic failures in large metal structures (Parker, 1957). The failures, which included ocean going freighters, gas pipe-lines, storage tanks and bridges, involved brittle fractures in nominally ductile metals. Contributors to the evolving theory which became known as 'fracture mechanics' were Orowan (1952), who generalized the work of Griffith to include small scale, plastic yielding, and Irwin (1948), who introduced the concept of the energy release rate which occurrs during crack extension.

The application of fracture mechanics requires the determination of a parameter which characterizes the stress state in the neighborhood of a crack tip. This parameter, which is called the stress intensity factor, must be determined from elasticity solutions which include a description of the crack geometry, the body geometry and the details of the applied loads. The required information can be generated by obtaining elasticity solutions which are valid at least in the vicinity of the crack tip.

The number of complete solutions to cracked body problems is limited by the need to impose relatively restrictive conditions on crack geometry, body geometry and loading. Test specimens used in the laboratory and components in structures and machines often involve configurations for which the stress intensity factors must be determined by numerical methods. Solution options which are available for determining stress intensity factors are described in section 2.4.3.

### 2.4.2 Crack modes

The cracks generated in service may, by virtue of part geometry and loading, be combinations of three distinct modes of crack extension. These are illustrated by the sketches in Fig. 2.6. These individual modes are described as the opening mode, the sliding mode and the tearing mode. The shaded areas represent the crack face surfaces.

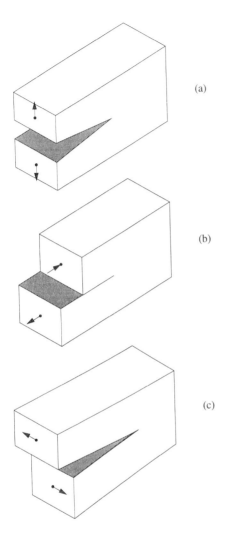

(a)

(b)

(c)

**Fig. 2.6** Crack opening modes

A combination of modes I and III is developed at the tip of a circumferential crack in a solid circular cylinder loaded by superimposed tensile and torsional loading. A crack which is inclined to the direction of loading in a plate will exhibit a combination of modes I and II. The tensioned plate in Fig. 2.7 illustrates this state of loading.

**Fig. 2.7**  Combined mode I and II loading

The occurrence of a more complex state of loading is illustrated in Fig. 2.8. The element shown represents a free body element cut from a structural component. With the boundary loading indicated an embedded semi-elliptical surface crack could exhibit all three crack modes. Modes I and II are developed at points A and B. At point C a combination of modes I and III would be indicated. It should be observed that in the absence of the opening mode in this example, friction between sliding fracture surfaces could be developed. Additional fracture surface forces can also be introduced by fracture surfaces that are not perfectly flat. The consequences of this type of behavior are examined in more detail in Chapter 5.

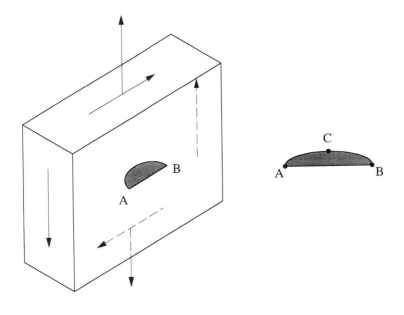

**Fig. 2.8**  A complex mixed mode state

### 2.4.3 Analysis of cracked bodies

The concepts embodied in fracture mechanics require the availability of solutions of the governing equations of the theory of elasticity. Two distinct approaches to the analytical solution of elasticity problems can be used. For a homogeneous, isotropic, linear elastic solid a displacement formulation can be generated by substituting Hooke's law (equation 2.26) into the equilibrium equations (equations 2.13, 2.14, and 2.15) to express them in terms of the strain components. The strain – displacement relations (equations 2.21 and 2.22) can then be used to obtain equilibrium equations in terms of displacements. Satisfaction of these equations and the boundary conditions provides solutions for both simply and multiply connected bodies.

A stress formulation can be obtained by substituting Hooke's law (equation 2.26) into the six compatibility equations. Satisfaction of these equations and the boundary conditions provides solutions for simply connected bodies. Satisfaction of additional relations are required to ensure single valued displacements in multiply connected bodies (Fung, 1965).

Crack problems of the type illustrated in Fig. 2.8 require the use of the three dimensional theory of elasticity. An introduction to the concepts of fracture mechanics is most easily initiated by a consideration of the simpler two dimensional cracked body problem.

For two dimensional problems the compatibility equation (equation 2.24) can, by use of Hooke's law and the equilibrium equations, be written as

$$\nabla^2(\sigma_{11}+\sigma_{22})=\left(\frac{\partial^2}{\partial x_{11}^2}+\frac{\partial^2}{\partial x_{22}^2}\right)(\sigma_{11}+\sigma_{22}). \tag{2.28}$$

Use of the Airy stress function, $\phi$, which is defined by the relations

$$\sigma_{11}=\frac{\partial^2\phi}{\partial x_{22}^2}, \quad \sigma_{22}=\frac{\partial^2\phi}{\partial x_{11}^2} \quad \text{and } \sigma_{12}=-\frac{\partial^2\phi}{\partial x_{11}\partial x_{22}}$$

converts equation (2.28) to the biharmonic equation

$$\nabla^2\nabla^2\phi=\nabla^4\phi=\frac{\partial^4\phi}{\partial x_{11}^4}+2\frac{\partial^4\phi}{\partial x_{11}^2\partial x_{22}^2}+\frac{\partial^4\phi}{\partial x_{22}^4}=0. \tag{2.29}$$

Note that the Airy stress functions identically satisfy the equilibrium equations.

For cracked body problems it is convenient to use polar coordinates. Transforming equation (2.29) from Cartesian to polar coordinates gives

$$\left(\frac{\partial^2}{\partial r^2}+\frac{1}{r}\frac{\partial}{\partial r}+\frac{1}{r^2}\frac{\partial^2}{\partial\theta^2}\right)\left(\frac{\partial^2\phi}{\partial r^2}+\frac{1}{r}\frac{\partial\phi}{\partial r}+\frac{1}{r^2}\frac{\partial^2\phi}{\partial\theta^2}\right)=0. \tag{2.30}$$

The Airy stress function in polar coordinates is defined by the relations

$$\sigma_{rr} = \frac{1}{r}\frac{\partial\phi}{\partial r} + \frac{1}{r^2}\frac{\partial^2\phi}{\partial\theta^2}$$

$$\sigma_{\theta\theta} = \frac{\partial^2\phi}{\partial r^2}$$  (2.31)

$$\sigma_{r\theta} = -\frac{\partial}{\partial r}\left(\frac{1}{r}\frac{\partial\phi}{\partial\theta}\right).$$

For the cracked body problem the solution function must be chosen to satisfy the boundary conditions on the traction free crack surfaces. The geometry for a semi-infinite crack in an infinite body is shown in Fig. 2.9.

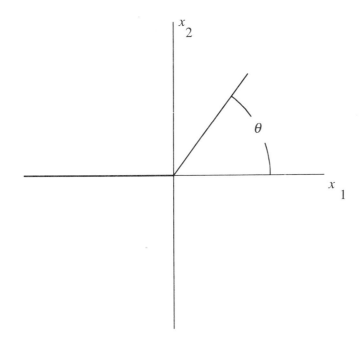

**Fig. 2.9**  A semi-infinite crack in an infinite body

Williams (1957) chose a solution function which can be used to examine the stress state developed in loaded, cracked bodies. The solution function which he used had the form

$$\phi = r^{\lambda+1}\big[C_1 \sin(\lambda+1)\theta + C_2 \cos(\lambda+1)\theta$$
$$+ C_3 \sin(\lambda-1)\theta + C_4 \cos(\lambda-1)\theta\big].\tag{2.32}$$

Equation (2.30), the governing equation, is homogeneous. Also, by virtue of the traction free crackface surfaces, the boundary condition equations for normal and shear stresses are homogeneous. The problem can, therefore, be formulated as an eigenvalue problem. The values of the parameters, $\lambda$, can be identified as being eigenvalues and each corresponding $\phi$ is then an associated eigenfuction. Note that the cosine and sine functions represent, respectively, symmetric and antisymmetric forms of $\phi$. The $n$th eigenfunction can be written as

$$\phi_m = r^{3/2}\Big\{ A_n\big[\cos(\tfrac{n}{2}+1)\theta - f_n\cos(\tfrac{n}{2}-1)\theta\big] + B_n\big[g_n\sin(\tfrac{n}{2}+1)\theta + \sin(\tfrac{n}{2}-1)\theta\big]\Big\}, \tag{2.33}$$

$$\text{where } f_n = \frac{n+2}{n+2(-1)^n} \quad \text{and} \quad g_n = \frac{2(-1)^n - n}{n+2}.$$

It follows then from equations (2.31) and (2.33) that the stress components for the symmetric mode I case are

$$\sigma_{rr} = \sum_{n=1}^{\infty} \tfrac{1}{4} nr^{\frac{n}{2}-1} A_n\big[-(n+2)\cos(\tfrac{n}{2}+1)\theta + (n-6)f_n\cos(\tfrac{n}{2}-1)\theta\big].$$

$$\sigma_{\theta\theta} = \sum_{n=1}^{\infty} \tfrac{1}{4} n(n+2)r^{\frac{n}{2}-1} A_n\big[\cos(\tfrac{n}{2}+1)\theta - f_n\cos(\tfrac{n}{2}-1)\theta\big]. \tag{2.34}$$

$$\sigma_{r\theta} = \sum_{n=1}^{\infty} \tfrac{1}{4} nr^{\frac{n}{2}-1} A_n\big[(n+2)\sin(\tfrac{n}{2}+1)\theta - (n-2)f_n\sin(\tfrac{n}{2}-1)\theta\big].$$

These equations for the stress components can be used as a basis for developing collocation solutions to crack problems. For a mode I problem, for example, the coefficients, $A_n$, can be evaluated by requiring that the stress components satisfy the outer boundary conditions at a discrete number of points. The resulting system of linear algebraic equations can then be solved for the $A_n$. Gross, Srawley and Brown (1964) used this procedure to obtain a solution of the problem of a tensioned sheet with an edge crack.

The character of the stress distribution developed can be revealed by examining the powers of $r$ in the sums for the stress components. The powers are:

$$-\tfrac{1}{2}, 0, \tfrac{1}{2}, 1, \tfrac{3}{2}, 2, \tfrac{5}{2}, \ldots$$

The first term has a singularity of the order of one-half and the strength of the singularity is determined by the coefficient of the first term. It may be concluded then that the character of the stress state immediately adjacent to the crack tip is indicated by the coefficient of the first term. It may also be concluded that all cracked bodies have the same singularity. The states differ by virtue of different coefficients for different crack and body geometries and conditions of loading. The parameter which has been chosen to reflect these differences has been called the stress intensity factor which for mode I loading is defined in equation (2.34) as

$$K_\mathrm{I} = 3\sqrt{2\pi}\,A_1 \tag{2.35}$$

The stress components can, therefore, be written as

$$\sigma_{rr} = \frac{K_\mathrm{I}}{4\sqrt{2\pi r}}\left(5\cos\tfrac{\theta}{2} - \cos\tfrac{3\theta}{2}\right) + \cdots$$

$$\sigma_{\theta\theta} = \frac{K_\mathrm{I}}{4\sqrt{2\pi r}}\left(3\cos\tfrac{\theta}{2} + \cos\tfrac{3\theta}{2}\right) + \cdots \tag{2.36}$$

$$\sigma_{r\theta} = \frac{K_\mathrm{I}}{4\sqrt{2\pi r}}\left(\sin\tfrac{\theta}{2} + \sin\tfrac{3\theta}{2}\right) + \cdots$$

$$\sigma_{zz} = v_1\left(\sigma_{rr} + \sigma_{\theta\theta}\right)$$

where $v_1 = 0$ for plane stress and $v_1$ is Poisson's ratio, $v$, for plane strain.

Equations for displacements $u_r$, $u_\theta$ and $u_z$ for $\phi_1$ are

$$u_r = \frac{K_I}{2E}\sqrt{\frac{r}{2\pi}}(1+v)\left[(2k-1)\cos\tfrac{\theta}{2}-\cos\tfrac{3\theta}{2}\right],$$

$$u_\theta = \frac{K_I}{2E}\sqrt{\frac{r}{2\pi}}(1+v)\left[-(2k-1)\sin\tfrac{\theta}{2}+\sin\tfrac{3\theta}{2}\right],$$

$$u_z = -\left(\frac{v_2 z}{E}\right)(\sigma_{rr}+\sigma_{\theta\theta}),\qquad\qquad\qquad (2.37)$$

where $k = \dfrac{(3-v)}{1+v}$, $v_1 = 0$ and $v_2 = v$      for plane stress,

and    $k = (3-4v)$, $v_1 = v$ and $v_2 = 0$      for plane strain.

The limitations on the use of the first term to characterize the mode I stress state at the crack tip and the roles of higher order terms are discussed in section 2.4.4 and in Chapter 5.

Equations for stress and displacement components for mode II can be obtained by use of the sin terms for $\phi_n$ in equation (2.32). The components resulting from the use of $\phi_1$ are as follows:

$$\sigma_{rr} = \frac{K_{II}}{\sqrt{2\pi r}}\sin\tfrac{\theta}{2}\left(1-3\sin^2\tfrac{\theta}{2}\right).$$

$$\sigma_{\theta\theta} = \frac{3K_{II}}{\sqrt{2\pi r}}\sin\tfrac{\theta}{2}\cos^2\tfrac{\theta}{2}.$$

$$\sigma_{r\theta} = \frac{K_{II}}{\sqrt{2\pi r}}\cos\tfrac{\theta}{2}\left(1-3\sin^2\tfrac{\theta}{2}\right).\qquad\qquad (2.38)$$

$$u_r = \frac{K_{II}}{2E}\sqrt{\frac{r}{2\pi}}(1+v)\left[-(2k-1)\sin\tfrac{\theta}{2}+3\sin\tfrac{3\theta}{2}\right].$$

$$u_\theta = \frac{K_{II}}{2E}\sqrt{\frac{r}{2\pi}}(1+v)\left[-(2k-1)\cos\tfrac{\theta}{2}+3\cos\tfrac{3\theta}{2}\right].$$

Similar relations can be obtained for mode III. These are summarized below:

$$\sigma_{rz} = \frac{K_{III}}{\sqrt{2\pi r}}\sin\tfrac{\theta}{2}.$$

$$\sigma_{\theta z} = \frac{K_{III}}{\sqrt{2\pi r}}\cos\tfrac{\theta}{2}.\qquad\qquad\qquad (2.39)$$

$$u_z = \frac{2K_{III}}{\mu}\sqrt{\frac{r}{2\pi}}\sin\tfrac{\theta}{2}.$$

Stress intensity factors can be obtained by use of the theory of complex variables. Analytical procedures developed by Muskhelishvili (1963) and by Westergaard (1939) form the basis for these methods. The procedures of Westergaard are described in Sih (1966) and by Pickard (1985). Introductions to the methods of Muskhelishvili are contained in Sokolnikoff (1956) and in Timoshenko and Goodier (1970).

The collocation method is not computationally efficient and the complex variable methods are not convenient for use in the types of cracked bodies encountered in laboratory specimens and in service components. These latter examples often also involve three dimensional cracks of the type illustrated in Fig. 2.8. For these more complex cases it has been necessary to resort to the use of the numerical procedures which make use of the theory of finite elements. Applications of finite element methods to cracked body problems are developed in Pickard (1985) and in Owen and Fawkes (1983). The latter reference also contains an introduction to the use of boundary elements. The application of boundary elements to fracture mechanics is also discussed in a report by Cruse (1975).

Collections of stress intensity factors for cracked bodies are included in the form of handbooks by Sih (1973a), Tada, Paris and Irwin (1973) and Rooke and Cartwright (1976). A collection of articles on analyses of crack problems is contained in a book edited by Sih (1973b). Introductions to fracture mechanics are presented in texts by Broek (1982), Hellen (1984) and Knott (1973).

Stress intensity factors for a few basic crack types and loading conditions are given in Appendix B.

### 2.4.4 Energy absorption during fracture

In a classic article Griffith (1920) determined the energy which was absorbed during the extension of a crack in a brittle solid. He identified the absorbed energy as that which was associated with the surface tension of the increment of the crack extension. The end result of his analysis was the conclusion that the stress necessary for crack extension is inversely proportional to the square root of the crack length, i.e.

$$\sigma_c \propto \sqrt{\frac{E_t}{a}} \ . \tag{2.40}$$

$\sigma_c$ is the critical applied tensile stress, $E$ is Young's modulus, $t$ is the surface energy per unit area, and $a$ is the crack length.

Equation (2.39) has been used successfully to estimate fracture stresses in brittle materials such as glass and ceramics. Under certain conditions, however, nominally ductile metals can fracture in a brittle manner. These conditions include low temperature, high strain rate and a localized geometric constraint which introduces a stress state which approaches that of a hydrostatic tension. Each of these conditions can lead to a suppression of plasticity. In recognition of this behavior Orowan (1952) suggested that equation (2.40) could be generalized to describe crack extension in cases in which the amount of plastic deformation which developed is small. Griffith's equation was thus modified to the form

$$\sigma \propto \sqrt{\frac{E(t+p)}{a}}, \tag{2.41}$$

where $p$ is the plastic deformation energy per unit area. For metallic solids $p >> t$.

Irwin (1948) in considering the brittle fracture of nominally ductile metals performed an analysis which established a relationship between the stress intensity factor and a quantity which he defined as the elastic energy release rate, $G$. In terms of the energy terms in equation (2.41) it follows that $G = 2(t + p)$. The change in potential energy, $\delta U$, during a mode one virtual crack extension may be written as

$$\delta U = G_I \, \delta a = \delta W - \delta V, \tag{2.42}$$

where $\delta W$ is the variation in external work and $\delta V$ is the variation in strain energy. Since it can be shown that the change in $\delta U$ which occurs during crack extension is independent of the external loading conditions (fixed grip versus constant load), it is possible to confine attention to energy changes in the immediate vicinity of the crack tip. Irwin (1957) showed that $G$ could be related to the stress intensity factor by computing the difference in energy between bodies with crack lengths of $a$ and $a + \delta a$. This can be accomplished by determining the energy required to close the upper crack in Fig. 2.10 to the configuration shown for the lower crack. The opening displacements from $r = 0$ to $r = a$ can be obtained from equation (2.37). Thus, for plane strain,

$$u_\theta(\delta a - r, \pi) = -\frac{4K_I}{E}\left(1 - v^2\right)\sqrt{\frac{\delta a - r}{2\pi}}\ ,$$ (2.43)

for $0 \le r \le \delta\ a$. The distribution of stress required to close the upper crack to the lower state is simply the distribution corresponding to the lower crack, i.e.

$$\sigma_\theta(r,0) = \frac{K_I}{\sqrt{2\pi r}}.$$ (2.44)

The change in energy is then

$$G_I\ \delta a = \lim \delta a \to 0 \int_0^{\delta a} \sigma_\theta(r,0) u_\theta(\delta a - r, \pi) dr.$$ (2.45)

Substitution of the terms of the Williams' series into equation (2.45) and performing the indicated operations gives

$$G_I = \frac{\left(1 - v^2\right)}{E} K_I^2$$ (2.46)

for plane strain. For plane stress

$$G_I = \frac{K_I^2}{E}.$$ (2.47)

By performing similar operations of the type indicated in equation (2.45), equations for energy release rates in terms of mode II and mode III stress intensity factors can be obtained. These are

$$G_{II} = \frac{\left(1 - v^2\right)}{E} K_{II}^2, \text{ and } G_{III} = \frac{(1 + v)}{E} K_{III}^2.$$

Problems in which two or three modes are present require the introduction of criteria which combine the energy release rates and the stress intensity factors in some functional form. The topic of mixed mode fatigue crack growth is discussed in section 5.6.

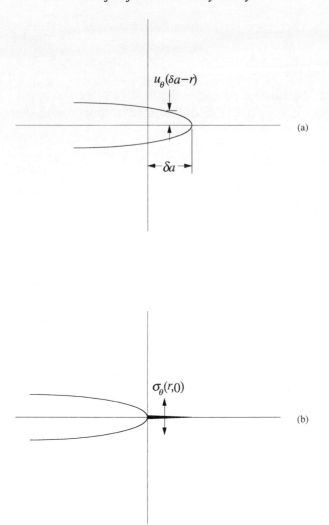

**Fig. 2.10**   Incremental closure of a crack tip

If mode I fracture occurs in a brittle manner, the values of the stress intensity factor and energy release rate are described as having reached their critical values, which are denoted as $K_{IC}$ and $G_{IC}$, respectively. Fracture toughness,$K_{IC}$ , is a mechanical property and values for various alloys have been determined and are tabulated in the literature. Tests for determining these values are subject to restrictions which ensure the validity of values obtained.  The restrictions are imposed to ensure that small scale yielding has occurred at the crack tip. Standards which have been developed require that the specimen thickness, $t$, and the crack length, $a$, both be equal to or greater than

$$2.5\left(\frac{K_{IC}}{\sigma_{ys}}\right)^2, \tag{2.48}$$

where $\sigma_{ys}$ is the yield strength. In addition certain requirements are imposed on the features of the load versus crack opening displacement curve. These details are described in standards developed by the American Society for Testing and Materials in ASTM Standard E399-74 (1974).

For the procedures developed for determining $K_{IC}$ it is assumed that the stress intensity factor provides a description of the mode I stress state at the crack tip for all geometries and loading conditions. Larsson and Carlsson (1973) have, however, shown that even for small scale yielding, plastic zone sizes determined by a boundary layer solution using $K$ can have significant errors. This was shown by comparing zone sizes determined by the boundary layer solution using $K$ with those determined by an elastic – plastic, finite element analysis. They further demonstrated that the use of the second, constant term of the Williams' series, when added to the singular term, provides acceptable approximations to the zone sizes determined by the elastic – plastic analyses. In their modified boundary layer solution the second term introduced a constant normal stress which is parallel to the crack plane.

This constant stress, which corresponds to the $n = 2$ term of equation (2.34), can be either tensile or compressive, depending upon the geometry of the cracked specimen. It was found to be positive for bend and compact tension specimens and negative for double edged and centre cracked specimens. In absolute magnitude it was smallest for their bend specimen.

The constant stress has been described as a $T$ stress. In Cartesian coordinates the total stress for the modified boundary layer solution can be written in matrix form as

$$\begin{pmatrix} \sigma_{xx} & \sigma_{xy} \\ \sigma_{yx} & \sigma_{yy} \end{pmatrix} = \frac{K_I}{\sqrt{r}}\begin{pmatrix} f_{xx(\theta)} & f_{xy(\theta)} \\ f_{yx(\theta)} & f_{yy(\theta)} \end{pmatrix} + \begin{pmatrix} T & 0 \\ 0 & 0 \end{pmatrix}. \tag{2.49}$$

Rice (1974) subsequently presented analytical results which supported the findings of Larsson and Carlsson. Leevers and Radon (1982) have provided results for several geometries and loading conditions and generated curves which can be used to determine the magnitudes of the constant stress term for different crack length to specimen width ratios. The results obtained by these investigators indicate that some differences in observed values of fracture toughness may be associated with geometric effects.

The above behavior includes the cases in which brittle fracture has occurred in nominally ductile materials. Usually these fractures occur under plane strain conditions in which plastic yielding is restricted. Methods for testing and analyzing elastic – plastic crack extension are discussed by Rice, Paris and Merkle (1973) and in British Standards BS 5762-1979 (1979).

Another aspect of brittle fracture which can be encountered may be revealed by reference to the integration indicated in equation (2.45). In performing the integration the energy release rate, $G$, was assumed to be constant. In a number of instances, however, microstructural variations with position and time can be developed, and then it can be anticipated that the fracture toughness, as measured by $G$, will also vary. Examples of this include base metal – weld zone regions, steels in which hardenability affects the depth of hardening, metals subject to elevated temperature embrittlement in environments with stress and temperature gradients, and metals exposed to radiation damage. In these examples the elastic properties are nominally unchanged, so elasticity theory still applies. For a fatigue crack traversing such regions, however, the conditions for unstable crack growth will vary with location. Thus, depending on the direction of crack growth, the effective value of $G$ may be either increasing or decreasing.

An analysis of the consequences of a crack advancing through a region with a fracture toughness gradient has been performed (Carlson, 1989). Since $G$ then becomes a function of position, it becomes necessary to include higher order terms of the Williams series to determine the conditions for crack advance. For a constant $G_c$ the corresponding critical stress intensity factor represents an unstable condition because crack extension can occur under a decreasing load.

For a decreasing fracture toughness, instability is increased. It can, however, be shown that for sufficient increases in fracture toughness an extending crack can be arrested (Carlson, 1989).

The gradient effects described may also be useful in providing a qualitative explanation for the behavior observed for fatigue and creep results for which the stress intensity factor provides a basis for correlating crack growth rate. For these cases, the applied stress intensity factors are less than the critical values and crack growth is described as stable. Adjacent to the crack tip, however, the effective $G$ values may be expected to be lower than those for the material outside the process zones. Crack advance may then occur when $G$ in the process zone decreases to a value corresponding to the applied stress intensity factor. If the crack then advances to a region of larger $G$, crack arrest could occur. Thus, the identification of the crack tip zone as a region with a toughness gradient may be used to provide a rational explanation for stable growth. This use of a modified form of equation (2.45) to account for incremental crack advance through an enclave with a toughness gradient is consistent with damage accumulation models which have been proposed by McClintock (1963), Weertman (1966) and Rice (1967).

The use of the stress intensity factor as a driving force for fatigue crack growth is discussed in Chapter 5.

# 3

# Macroscopic and microscopic features of fatigue

## 3.1 INTRODUCTION

Pook (1983) has observed that providing a rigorous, satisfactory definition of metal fatigue is difficult. A variety of definitions have been offered, but the elements of most are similiar. Many texts on fatigue have dealt with metals and the discussions have, accordingly, focused on mechanisms which are operative in metals (Fuchs and Stephens, 1980; Bannantine, Comer and Handrock, 1990). An exception to this is a book by Suresh (1991) which covers both metals and nonmetals. A typical definition which has been offered by Bannantine, Comer and Handrock (1990) is: 'metal fatigue is a process which causes premature failure or damage to a component subjected to repeated loading'. Although this definition is satisfactory, it could be made more general. As noted in section 1.3, fatigue has been found to be not exclusively limited to metals. Replacing the word 'metals' by 'solids' would recognize this development. It may also be noted that although fatigue can lead ultimately to failure, the basic feature which is common in the behavior of all of the reported cases is the stable, as opposed to catastrophic, growth of cracks.

The mechanisms which lead to the formation and growth of cracks differ, however, for different types of solids. In view of the fundamental differences in atomic bonding and microstructure such differences may, of course, be anticipated. To focus attention on the features which these solids have in common, however, the following definition may be considered:

Fatigue in solids is a phenomenon in which repeated or varying load creates damage in the form of cracks which can grow in a stable manner.

Note that since fatigue can result in the formation of arrested cracks, it is not necessary to include failure as a fatigue event.

As noted in the Chapter 1, engineers very early identified fatigue as a design problem and established measures to guard against it without knowing very much about operative mechanisms. Empirical, rational design procedures were developed, and these types of approaches still characterize existing practices. The use of damage tolerance procedures and the introduction of a wide variety of materials has, however, increased the importance of being able to identify and categorize the primary mechanisms that are operative in each type of application. A recognition of this importance can contribute to improvements in engineering design.

To emphasize here the differences in fatigue mechanisms for different solids, each type of solid for which fatigue has been reported is discussed in a separate section. The first three sections cover metals, plastics and brittle solids. In view of the expanding interest in the potential uses of manufactured composite materials, the final three sections are on composites with metallic, polymeric and ceramic matrices.

## 3.2 METALS

Pook (1983) has estimated that during the period between 1838 and 1983 about 20000 papers were published on the subject of fatigue. The vast majority of these papers reported results of investigations of metal fatigue. Prior to about 1955 nearly all fatigue tests were conducted on what have often been described as 'smooth specimens'. Bar or component specimens were tested as fabricated and the introduction of pre-existing cracks was not considered a part of the test procedure. This discussion of fatigue in metals begins, therefore, with a consideration of the mechanisms which are operative in 'smooth specimen' type tests.

Fatigue damage which results in the nucleation of cracks can be initiated in a layer at a free surface. It can occur as a result of localized, irreversible slip in narrow bands along slip planes in individual grains. Since the band region terminates at a free, unconstrained surface, roughness in the form of extruded material can be created. A fatigue process in which layers within slipbands irreversibly slide relative to one another has been proposed by Forsyth (1953) as an explanation for the appearance of surface irregularities in the form of 'extrusions' and 'intrusions'. High concentrations of plastic strain are developed within these localized bands and they can ultimately serve as sites for the initiation of microcracks.

It has been shown that if a surface layer containing fatigue damage is removed by polishing, there is an improvement in fatigue life (Alden and Backofen, 1961). This clearly focuses attention on the importance of the condition of the surface layer. It reveals that fatigue behavior can be expected to be susceptible to treatments and exposures which alter the condition of the surface.

Comparisons of results from experiments conducted in active and inert environments have also demonstrated the sensitivity of fatigue behavior to surface conditions. Formation of an oxide on partially exposed slip surfaces can, for example, inhibit reverse slip during load cycling and thereby promote surface roughening (Thompson, Wadsworth and Louat, 1956). The presence of an active environment can, therefore, contribute to a decrease in resistance to crack nucleation by virtue of an increase in surface roughness. Aggressive atmospheres which can produce cracks even without external loading can, with loading, result in stress – corrosion cracking. This topic is discussed in section 5.7.4.

Much of the accumulated knowledge on fatigue crack nucleation has been the result of research investigations on high purity metals. Commercial alloys used in structural systems and machines are not subject to the rigid controls used in research, and a variety of manufacturing and operational variables can affect the crack nucleation process.

Manufacturing metallic components involves the production of alloys and the fabrication of structural elements. Potential sites for crack initiation include voids and brittle inclusions which may be cracked prior to or shortly after a component is introduced into service. These, when located in a surface layer, can act as stress raisers which serve as slip initiation sites. Blom *et al*. (1986) found that crack nucleation under fatigue loading in the aluminum alloy 2024-T3 initiated at intermetallic particles. In the aluminum alloy 7475-T761, however, they found that crack nucleation initiated both along particle/matrix interfaces and at slipbands within grains. Swain *et al* . (1988) traced the initiation of fatigue cracks in a 4340 steel to inclusions at or just below the surface of test specimens.

A variety of processes are used to convert metals into forms which can be fabricated into components. These include forging, casting, extruding, drawing and rolling. Each of these processes has inherent features which can be sources of flaws which serve as sites for the initiation of fatigue cracks.

The fabrication of components from the many forms of stock available usually requires some machining operations which create new surfaces. The properties of the resulting surface layers depend on the type of machining operation used. These operations can produce disturbed surface layers which introduce residual stresses. Residual compressive stresses can inhibit slip and delay the nucleation of cracks. Grinding and shot peening surfaces introduce residual compressive stresses and can result in endurance limits which are substantially greater than those for polished, stress free surfaces. This is illustrated in Fig. 3.1. Metallurgical processes such as nitriding, carburizing, and flame and induction hardening create high strength surface layers which have increased resistance to slip.

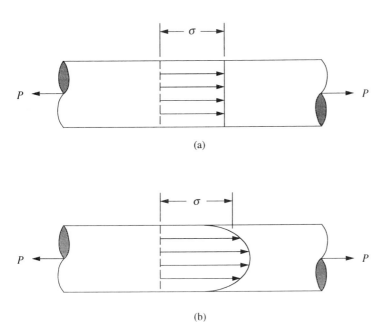

**Fig. 3.1** Bar stress distributions with and without residual stress for (a) applied tensile stress only and (b) applied tensile stress with compressive surface residual stress

Surface conditions can also be altered by flaws which are introduced during assembly and servicing. Improper use of tools can introduce scratches and gouges which can, when located at critical stress locations, result in potential sites for crack nucleation. Failure reports often identify such flaws as the initiation site for fatigue failures.

A large portion of the total time to failure, as depicted in Fig. 1.1, is consumed during the crack nucleation phase. Since the nucleation mechanics can vary significantly from one specimen to another, considerable data scatter can occur in stress versus cycles to failure tests. Thus, at a given stress level it becomes necessary to consider the use of statistics and deduce the probability of failure at a given number of cycles. This is qualitatively illustrated in Fig. 3.2. Note that scatter is largest at the low stress region where stresses approach what is defined as the endurance limit. Statistical estimates of confidence in test results may be obtained by the use of analyses of variance. The basic elements of fatigue testing and the statistical analysis of fatigue data are discussed by Weibull (1961) and Johnson (1964). The use of statistical theory in the management of aircraft fleets is discussed in Chapter 6.

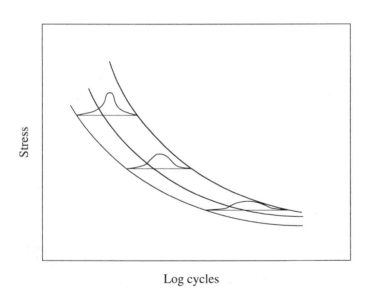

Fig. 3.2   Statistical interpretation of fatigue scatter

Beginning in about 1955, there was a change of emphasis in fatigue testing. Instead of conducting only smooth specimen tests, researchers began performing experiments on pre-cracked specimens. In these experiments the crack lengths are monitored and crack growth histories are obtained. For the same specimen geometry and loading conditions, it has been found that the reproducibility of these test data is very good (Pook, 1983). This is in sharp contrast to the scatter behavior observed for stress versus cycles to failure tests on smooth specimens.

As a crack advances under cyclic load, a cyclic plastic zone is formed in front of the crack tip. A parameter which characterizes the stress state at a crack tip is defined as the stress intensity factor. In Chapter 5 it is shown that the range of the stress intensity factor is proportional to the square root of the cyclic plastic zone. It has been found that the rate of fatigue crack growth can be expressed as a function of the range of the stress intensity factor. It follows that the use of the square root of the cyclic plastic zone size provides an equivalent growth rate parameter. Although the use of the stress intensity factor has distinct advantages in analyses of crack growth in components, it is easier at this stage of the discussion to visualize the relation between growth rate and cyclic plastic zone size. It should be noted, moreover, that the two parameters are not equivalent for all conditions. If crack tip yielding exceeds certain limitations (small scale yielding), the customary use of the linear theory of elasticity must be re-examined, because the stress intensity factor no longer characterizes the state at the crack tip. A cyclic plastic zone is, nevertheless, developed and it can still be identified as the fatigue process zone in which crack growth evolves. Its size and shape must, however, be determined by different analytical methods. This limitation is discussed in Chapter 5.

In this section, the stages of fatigue crack growth rate are described in terms of the size of the cyclic plastic zone. Under fixed cyclic loading conditions a fatigue crack will lengthen and the size of the cyclic plastic zone will increase. The resulting behavior can be characterized by use of Fig. 3.3, which is a plot of the log of the crack growth rate, $da/dN$, versus the log of the square root of the cyclic plastic zone, $\rho_{cyc}$. The curve of Fig. 3.3 provides a qualitative description of features which are typical for fatigue crack growth in metals.

The growth rate behavior in region II, which is described as the Paris region, is linear and this implies that a power relationship between the rate of growth and the square root of the cyclic plastic zone is applicable. In this region cracks have been observed to grow by the

formation of parallel striations which range in size from 0.05 to 2.5 µm
(Bates and Clark, 1969). Tests have indicated that crack advance
sometimes occurs by the formation of a striation during each cycle of
loading with the striation size being determined by the magnitude of the
cyclic load. These observations can sometimes be used to deduce the
loading conditions which preceded a fatigue failure in service.

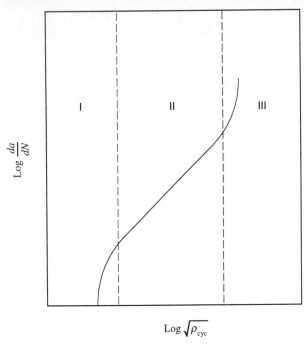

**Fig. 3.3** Rate of fatigue crack growth versus cyclic plastic zone size

Forsyth (1962) proposed a model which suggested that striation
formation was the result of a cleavage fracture ahead of the crack tip.
The subsequent rupture of the ligament which was formed was then said
to produce the observed striations. Laird and Smith (1962) proposed a
model which suggested that the striations were a result of alternate,
plastic blunting and sharpening of the crack tip during cyclic loading.
Secondary crack advance mechanisms have been observed. These
include abrupt, brittle intergranular or transgranular microfractures
which result in discrete growth increments (Ritchie and Knott, 1973,
Beevers *et al.*, 1975). The growth rate in region II has been described as

a continuum process and has not been found to be strongly dependent on microstructure.

Region I is characterized by a rapid decrease in crack growth rate with decreasing cyclic plastic zone size. Below the indicated threshold, growth rates of the order of $10^{-8}$ mm/cycle or less occur. The behavior observed within region I has been attributed to two forms of resistance to crack growth. These have been described as extrinsic resistance and intrinsic resistance (Beevers and Carlson, 1986). As a crack advances, an inelastically deformed fracture surface is formed and it creates an obstacle to complete crack closure during unloading. The obstacle to closure can be present in the form of an inelastically deformed layer and discrete asperities which are associated with fracture surface roughness. If complete closure is prevented, the effective driving force for crack advance is reduced. Whether or not the surface obstructions come in contact with one another during unloading depends upon the magnitude of the minimum applied stress. A measure of this effect is usually expressed in terms of the mean applied stress, i.e. for a given applied stress range, increasing the mean stress reduces the effect of closure prevention. The prevention of closure by crack surface obstacles introduces what has been described as an extrinsic resistance.

The second component of resistance could be deduced if obstructions to closure could be excised. The behavior which would then be developed would be due to the intrinsic resistance. Experimental results have indicated that crack growth resistance in region I increases with increasing stiffness and strength for metals and alloys which exhibit a dominantly transgranular mode of fracture (Beevers and Carlson, 1986). Other factors upon which behavior is dependent include slip character and crack branching, grain size, texture and environmental sensitivity.

Since the loading conditions encountered within region I usually lead to closure obstruction during unloading, the total resistance to crack growth is generally the sum of the intrinsic and extrinsic resistances. Both the processes which are involved in intrinsic resistance and the mechanisms associated with the formation of surface obstruction to closure are microstructure sensitive. (It should also be noted that the form of the closure obstruction is dependent on specimen thickness. This issue is discussed in Chapter 5). The behavior in region I is, therefore, dependent on microstructural features. Since the near threshold region is of considerable importance in service components, there are strong incentives for using metallurgical processes that produce microstructures which improve crack growth resistance.

The rate of growth in region III increases rapidly and terminates with fracture as *da/dN* increases without bound. Within this region the deformation mechanisms resemble those encountered in monotonic loading. These mechanisms, which result in a fibrous texture, cleavage and intergranular fracture, are strongly dependent on microstructure, mean stress and specimen thickness.

Active environments together with cyclic loading are often encountered in service components, and the resulting behavior is described as corrosion fatigue. This is a particularly important issue in aircraft structures by virtue of the consequences of failure. The effects of environment are most severe in region I. The coupled effects on crack growth in region II are mild for dilute environments. Environment has little effect in region III crack growth. Detailed descriptions of operative mechanisms in the three regions for specific alloys are contained in an article by Knott (1986).

## 3.3 POLYMERIC SOLIDS

Polymeric solids exist in a variety of forms. Their use as matrices in composites is discussed separately in section 3.6. The bonding forces in metals and polymeric solids are quite distinct. Metallic elements release electrons to develop metallic bonding. Nonmetallic elements which share electrons and develop covalent bonds form the basis for the development of polymeric solids.

Pure polymeric solids are formed by addition or condensation processes in which long molecular chains are created (Hertzberg, 1989). The chains in the resulting solids are quite mobile because bonding between chains is weak. These polymeric solids are described as being thermoplastic.

Plastics are polymeric materials with additives which modify both the structure and the properties of the original polymer. Among the additives which are used are those which create crosslinking between molecular chains. This process results in the formation of a strongly bound three dimensional network of polymer chains.

Themoplastic solids are ductile and with sufficient heating and applied stress they are subject to time dependent deformation or creep. Also, unlike thermosetting solids, they can melt.

Thermosetting solids tend to be brittle, do not melt and are generally much less sensitive to temperature induced creep than are thermoplastic solids.

Thermoplastic matricies which are commonly used in composites include polypropylene, nylon and polycarbonate. Polypropylene and nylon are usually semicrystalline, whereas polycarbonate is amorphous. Epoxy and polyester resins are often used to form thermosetting matrices for composites.

In spite of the fundamental differences between metals and plastics the responses to applied loads appear to be phenomenologically similar. It has been found that some polymeric solids exhibit stress versus cycles to failure curves which are similar to those for metals. Also, a stress below which failure does not occur has been observed (Hertzberg, 1989).

Three irreversible damage mechanisms which can occur in polymeric solids are crazing, slipband formation, and thermal degradation. Damage introduced by crazing and slipband formation can accumulate under either monotonic or cyclic loading.

In appearance crazing in polymers resembles that which occurs in pottery glazes. Whereas crazing in glazes results in a network of fine cracks, the upper and lower surfaces of what appear to be cracks in polymers are actually interconnected by systems of discrete fibrils. Also, in polymers the crazes are not restricted to an outer surface, but can extend into the interior. The density of the material in the craze zone ranges from 40 to 60% of that of the matrix material (Kambour, 1964).

Slipband formation can occur by shear yielding. Crack extension can proceed by a shearing process and by fracture through a craze zone. Takemori (1982) has observed the effects of interactions between these two processes. He observed the formation a periodic sequence of slipbands superimposed on a crack advancing through a craze zone.

Damage resulting from thermal degradation is an important factor when cyclic loading is involved. Above a 'glass transition' temperature, deformation in polymeric solids becomes time-dependent or viscoelastic. Energy which is absorbed under cyclic loading is transformed to heat. This process is described as hysteretic heating, and it reduces resistance to deformation as the weak bonds between molecular chains are further weakened. In thermosetting solids with cross-linking between chains the mobility is not significantly increased with increasing temperature, so the relaxation behavior which is exhibited by thermoplastic solids is suppressed. The decrease in fatigue strength in polymeric solids with increasing frequency is illustrated in Fig. 3.4. Internal heating with heat loss through an external surface is a problem in heat transfer. A consideration of the features of the problem reveals that the increase in temperature in a given time period depends on the surface to volume ratio. Thus, for two smooth bars with circular cross-sections and the same length, the bar with the larger diameter will undergo a greater increase in temperature. The larger bar will, consequently, suffer greater thermal degradation and exhibit a smaller resistance to fatigue.

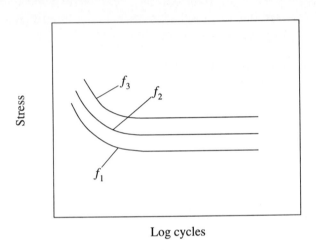

Log cycles

**Fig. 3.4**  Effect of cyclic frequency, $f$, on fatigue strength ($f_1 > f_2 > f_3$)

The preceding example was based on the application of a uniform stress, so the heat generated per unit volume would be expected to be uniform. The local high stresses generated adjacent to a stress concentration, such as a notch or a hole, will, under cyclic loading, generate local temperature gradients. Heat then is lost not only through the external surface, but also into the interior. The mechanics of the localized thermal degradation and damage will, consequently, differ substantially from that which occurs in a uniformly stressed body.

Increases in temperature are not only dependent on cyclic frequency and specimen size. Sauer and Richardson (1980) have conducted fatigue tests on two polymeric materials which have similar elastic moduli and yield strength. For the same cyclic frequency and applied loading the temperature of one – polystyrene – rose only a few degrees. The second polymer – polymethyl methacrylate – melted rapidly. They attributed the difference in behavior to the fact that viscoelastic damping in polymethyl methacrylate is much greater. Damping capacity must, therefore, be included as a factor which affects fatigue behavior.

Fatigue crack growth tests have also been conducted on polymeric solids (Hertzberg, Skibo and Manson, 1979). Correlations of the type used for crack growth in metals have been applied, but again the growth rates can be cycle frequency dependent. In contrast to smooth bar fatigue behavior, however, the crack growth rate in some polymeric materials decreases with increasing frequency. It has been suggested that the observed behavior stems from the viscoelasticity of polymeric solids, and that the crack growth rate at low frequencies is a coupled response involving both fatigue and time-dependent inelastic deformation or creep. At higher frequencies creep effects are decreased.

The effect of cross-linking on crack growth has been demonstrated by Hertzberg, Manson and Skibo (1975). They conducted fatigue crack growth tests on polystyrene and cross-linked polystyrene specimens. It was found that cross-linking both increased resistance to crack growth and reduced susceptibility to frequency effects. These results indicate that decreasing the mobility of the molecular chains reduces viscoelastic effects.

It is also of interest to note that Hertzberg, Skibo and Mason (1979) have observed striation markings in several amorphous polymeric solids. They found that each striation increment corresponded to a single load cycle. This indicates that growth by a striation mechanism can occur in both crystalline solids (metals) and in amorphous solids.

The fatigue behavior of polymeric solids has been comprehensively treated in a review article by Takemori (1984) and in books by Hertzberg and Manson (1980), Kinloch and Young (1983) and Kausch (1987).

## 3.4 BRITTLE SOLIDS

Ceramics, which are compounds of metallic and nonmetallic elements, include such solids as oxides, refractory materials, glass and concrete. Many have a crystalline structure and most have ionic or covalent bonding. Ceramic compounds such as refractory carbides and nitrides are exceptions and they have combinations of metallic and covalent bonds. Low fracture strength and negligible ductility are characteristic of these solids and they are commonly described as being brittle (Van Vlack, 1985, Kingery, 1960).

Many of the ceramic materials that are of potential interest in special engineering applications are fabricated by compacting fine particles or powders. They are then sintered by heating to agglomerate the particles into solids (Norton, 1974). The agglomerization process requires the formation of bonding between particles during elevated temperature sintering. Bonds are developed by two processes: formation of liquid phase interfaces which bond particles together; solid diffusion between particles which reduces internal free surfaces by elimination of gaps between particles (Van Vlack, 1985). The objective of the fabrication process is to develop, as nearly as possible, theoretical densities. The degree to which this is achieved is a measure of the success in eliminating pores or voids.

The possible uses of ceramics are limited by their inherent brittleness, i.e. their inability to tolerate under loading the presence of a crack without failing in a catastrophic manner. Some improvements in strength have been realized by the introduction of surface compressive stresses after quenching from high temperatures and by applying coatings with low thermal expansion glazes (Kirchner, Gruver and Walker, 1968). Also, mechanisms which may partially alleviate this intolerance have been discovered. These have been described as crack tip shielding mechanisms because they may reduce the intensity of crack tip loading which is produced by externally applied loading.

One of the crack tip shielding mechanisms is associated with the observed formation of microcracks in the neighborhood of a crack tip. Although there may be rational support for benefits of this behavior, it may also be argued that extensive microcracking could result in a degradation of toughness (Ortiz, 1988).

The results of a number of analytical studies have been published on this subject (Charalambides and McMeeking, 1987; Hutchinson 1987; Ortiz 1987; Kachanov, 1988). Experimental verification of predictions of the effects of microcracking are difficult to obtain, however, because of the localized character of the mechanism. Since direct, supporting evidence is not available, the benefits of this mechanism are not yet well enough established to be exploited in engineering design.

A second crack tip shielding mechanism involves a stress induced transformation. Garvie, Hannick and Pescoe (1975) observed that a stress induced transformation of metastable zirconia in ceramics results in a toughening response which resembles that found for metals. Marshall (1986) has shown that the loading and unloading characteristics of zirconia which is partially stablized with magnesia exhibit features similar to the elastic – plastic behavior observed for metals. The deviation, during loading, from linear-elastic behavior marks the beginning of a transformation from a metastable, tetragonal zirconia structure to a monoclinic structure. Unloading, like that for metals, is linear elastic. As with metals, this irreversible behavior denotes a toughness which is not normally observed for ceramics. Claussen (1981) has exploited this mechanism by using surface grinding to introduce compressive surface stresses in transformation toughened zirconia.

The potential toughening behavior exhibited by this class of ceramics has created interest with regard to its impact on crack growth. A crack tip region is highly stressed and the development of a localized, stress induced transformation zone is analogous to the crack tip plasticity zone in metals. Analyses which model this behavior have been performed (McMeeking and Evans, 1982; Budiansky, Hutchinson and Lambropoulos, 1983). Although this class of ceramic materials exhibits an enhanced toughness, it may be concluded that its use may initially be limited to providing improved replacement components in applications which already use more traditional ceramics.

The use of ceramics is of special interest in high temperature applications. Alumina, for example, is not subject to oxidation as are metals. Metals such as tungsten and niobium have very high melting temperatures, but they are subject to devastating oxidation in air atmospheres. Also, ceramics have been observed to exhibit inter-granular creep at high temperatures and this behavior can, by stress redistribution, alleviate the disadvantages of the brittleness of these materials. Intergranular creep can be expected to be most effective in reducing brittleness in those ceramics for which liquid phase bonding is developed during sintering. The resistance to shear in the viscous interfaces decreases with increasing temperature. The properties attendant with this type of bonding thus provide an opportunity to select ceramics which are best suited for elevated temperature applications.

Fully reversed loading fatigue tests have been conducted by Guiu (1978) on smooth specimens of alumina. As would be expected for fatigue, the cycles to failure increased with decreasing cyclic stress. Guiu compared these results with data from constant load tests (static fatigue). This comparison indicated that the application of the cyclic loading reduced the resistance to failure. Evans (1980) has suggested that observed fatigue failures in some ceramics may be a consequence of stress corrosion. If a corrosive environment is present, it could be expected that cyclic loading could add to progressive damage. During cyclic loading, the reversal in the direction of frictional forces acting on microcrack surfaces could conceivably also contribute more damage than a constant load.

Crack initiation and growth in brittle, notched specimens under cyclic compression has also been reported (Ewart and Suresh, 1987). Experimental evidence was offered to show that during cyclic loading, a network of microcracks develops in front of the notch tip. Growth normal to the axis of loading was then observed to occur by the coalescence of the microcracks. It was suggested that the primary cause of continuing, stable crack growth is the development of residual tensile stresses within the process zone.

To understand the process in which the initial cracks are developed, it may be observed that the specimen configuation used by Ewart and Suresh (1987) possesses a feature which smooth bars do not have, i.e. a notch which introduces a stress concentration. To understand the effect which kinematic constraint of the type introduced by such a notch can have, it is instructive to consider another type of loading condition. It has long been known that a brittle circular cylinder which is compressed between two anvils will sustain progressive damage in which rough, partial cone-like fracture surfaces are developed (Nadai, 1931). The fracture process, which begins at the rim contacting the anvil, has been attributed to anvil constraint which inhibits specimen expansion at the base. This behavior is illustrated in Fig. 3.5.

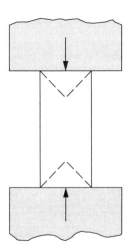

**Fig. 3.5**  Fracture in a brittle cylinder under compressive loading

It has also been shown that as the cylinder height is reduced, as shown in Fig. 3.6, the upper and lower fracture surfaces approach one another, the angles between the cone elements and the anvil surfaces decrease, and a very localized damage zone is developed. From these observations, it may be concluded that a notch with a small radius could focus the cracking into a very narrow band.

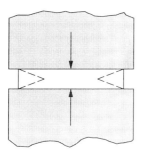

**Fig. 3.6**  Effect of cylinder height to diameter ratio on fracture

The damage incurred is, of course, irreversible, and if cyclic compressive loading were applied, the internal fracture surface gaps and misfits would be subjected to damage by reversed, frictional sliding contact. For repeated cycles of loading this could result in a progressive inward advance of damage in the form of crushed debris. Also, Ewart and Suresh (1987) found that the rate of advance of the damaged zone could be increased by ultrasonically removing the crushed debris. Although this has been described as stable crack growth, it may be more definitive to characterize it as a progressive, highly localized comminution.

Kishimoto *et al.* (1992) have examined the influence of a bridging mechanism on crack growth under tensile loading on alumina. They found that the crack path is not flat but consists of connected short segments or facets which are inclined to one another. During loading, the inclined upper and lower surfaces slide relative to one another, and produce localized contact forces and frictional forces which effectively introduce a bridging action. The bridging action was, however, found to be more effective for monotonic loading than for cyclic loading. Under cyclic loading, the interference between asperities results in microfractures which progressively reduce the obstructions to closure and hence the effectiveness of the bridging action.

An analytical model for the interaction between inclined fracture surface asperities has been developed by Carlson and Beevers (1985). The results indicate that the inclination angle of facet steps and the friction coefficient determine the local values of mode I and mode II stress intensity factors which are developed. Because of the presence of friction, the loading and unloading paths, during cycling, differ. The mixed mode behavior developed by the global and local loading produces a nonproportional loading state at the crack tip. For unloading, for example, $K_I$ decreases and $K_{II}$ increases during closure contact. The mixed mode conditions developed at the crack tip could, therefore, promote crack deflection and thereby reduce crack growth rates.

Cardona, Bowen and Beevers (1992) conducted positive $R$ ratio fatigue crack tests on partially stabilized zirconia. They attributed the crack growth resistance which they observed to the development of compressive residual stresses adjacent to the crack tip. To examine the influence of a transformation mechanism, a test was interrupted and the specimen was annealed to relieve the compressive residual stresses in the tranformed zone. Subsequent resumption of testing revealed that crack growth could occur at substantially lower loading levels. They cited this result as evidence of the effectiveness of the phase transformation in increasing crack growth resistance.

Reece, Guiu and Sammur (1989) have conducted fully reversed loading, crack growth tests on alumina specimens and have correlated their data by use of the Paris relationship. Dauskart, Marshall and Ritchie (1990) have reported data for tensile cyclic loading on the transformation-toughened zirconia, and their data have also been correlated by use of the Paris relationship. Elevated temperature fatigue crack growth tests at 1050 °C under cyclic tension have also been conducted by Ewart (1990) and Suresh (1990) on alumina specimens. These also have been correlated by use of the Paris relationship. Although there has been debate about the operative mechanisms for the above results (Suresh, 1991), attention should be focused on the values of the exponents found for the Paris relationship. For the room temperature data on alumina the exponent has been reported as $14 \pm 5$ (Suresh, 1991). For the zirconia the values ranged from 21 to 42. For elevated temperature tests on alumina, the exponents ranged from 8 to 10. These values indicate that either a very small change in load or a small amount of uncertainty due to scatter can result in very large changes in crack growth rate. Metals, by contrast, have exponents which are of the order of 2 to 3. The fatigue crack growth results which have been reported would then indicate that these materials could not be classified as damage tolerant.

The load bearing use of brittle solids is primarily confined to components in which compressive stresses are developed. The results which have been described indicate that the presence of stress concentrations, even under compressive loading, may be a component feature which should be avoided. Thus, although a few general recommendations can be specified, a number of issues remain to be resolved before sound design procedures can be established for this class of solids.

Another issue concerns pre-existing flaw distributions and how they affect size and stress distribution effects. Brittle fracture test results can be subject to considerable scatter and, as a consequence, statistical analyses have often been used to describe the behavior. Weibull (1939) developed a theory in which pre-existing distributions of flaws were present and the cases of both volume and surface flaws were considered. He used extreme value distributions to predict the basic features of the behavior and his results indicated that the fracture of brittle solids can be expected to be both specimen size and stress state dependent. Epstein (1948) has presented a thorough exposition of the application of the statistics of extreme value distributions to the problem of fracture.

Using a volume flaw distribution function, Weibull's analysis gives a relation for the ratio of the probable tensile strength for two bars of different size of the form

$$\frac{\sigma_1(\text{ult.})}{\sigma_2(\text{ult.})} = \left(\frac{V_2}{V_1}\right)^{\frac{1}{m}}, \tag{3.1}$$

where the $\sigma_i$ (ult.) values are tensile strengths for specimens with volumes $V_1$ and $V_2$. $m$ is a material constant greater than unity. This ratio predicts the existence of a size effect for which smaller bars would, on the average, have larger strengths.

For two bars of the same size with one loaded in tension and one in bending, Weibull's analysis gives the ratio

$$\frac{\sigma(\text{bend. ult.})}{\sigma(\text{tens. ult.})} = (2m+2)^{\frac{1}{m}}. \tag{3.2}$$

This reveals the effect of stress distribution. In bending only a small portion of the volume is exposed to high tensile stresses, so the bending specimens would tend to exhibit higher strength.

The above relationships, which indicate size and stress distribution effects in brittle materials, have been experimentally evaluated by Davidenkov (1947). Using brittle, high phosphorus steel specimens, he obtained results which give good, quantitative support for the basic concepts of Weibull's theory of brittle fracture.

Salmassy, Duckworth and Schwope (1955) and Salmassy, Bodine and Manning (1955), using plaster of Paris as a test material, conducted extensive tests on the effects of size and stress state. They obtained Weibull flaw distribution functions for their test material, and their results provide additional evidence for the basic validity of Weibull's theory.

The statistical analysis of brittle fracture is based on the proposition that the variation in the severity of initial flaws can be represented by a frequency distribution function. Under monotonic loading, the achievable strength then depends upon the stress state and the flaw distribution for the given body. If the initial loading is not critical and cyclic loading ensues, the flaw distribution is amended if noncritical microcracking occurs. During cycling, the flaw distribution, which consists of initial voids and processing imperfections and microcracks due to loading, is thus continuously being revised. Lamon (1992) has developed a statistical method of analysis which is designed to track the evolution of these distributions. The objective is to provide residual strength probabilities in terms of the number of cycles of loading. To illustrate the statistical basis of the progressive degradation process, he has used the results of thermal fatigue tests on silicon nitride specimens. Residual strength probability distributions for several values of thermal cycling were then deduced. The probability of failure versus strength curves for cycled specimens was substantially higher than those for as-received specimens.

## 3.5 METALLIC MATRIX COMPOSITES

Efforts to fabricate useful metallic matrix composites have been confronted with the necessity of solving a variety of technological problems. Quite different approaches have been adopted and the resulting fabrication techniques which have been developed are correspondingly distinct from one another. Metal matrix composites have been reinforced by use of particles, films and fibers.

### 3.5.1 Fabrication

Dispersion hardening has long been known to provide dramatic increases in strength by virtue of distributions of fine dispersoids which inhibit inelastic slip. Heat treatable aluminum alloys exploit this mechanism (Guy, 1962). Particle reinforced composites have much higher volume fractions of dispersoids than the commonly encountered commercial metals which are classified as dispersion hardened alloys. Also, the size of the particles in the metallic composites is much greater. At elevated temperatures, the finer particles in metallic alloys dissolve, and no longer provide dispersion strengthening. By contrast, particles of alumina, for example, in aluminum have low solubility and the benefits of hardening can be retained at higher temperatures.

Particulate reinforcements in metal matrices have been achieved by the use of molten metal mixing, powder metallurgy, and spray deposition. Stock for components can then be manufactured by extrusion, rolling or drawing.

Laminated composites can also be fabricated by repeated chemical vapor deposition of microscopic films on metallic matrix lamellae. In these composites the lamellar films provide high stiffness and strength and the metal matrix provides ductility and formability (Kreider, 1974). Another example of this class of composites involves eutectic compositions in which two phases solidify in a lamellar sequence. Those composites consist of a ductile metallic matrix which is reinforced by a stronger, stiffer lamellar phase (Thompson and Lempkey, 1974).

Considerable effort has been concentrated on the development of fiber reinforced metallic composites. Particular emphasis has been directed toward the development of systems having comparatively low strength matrices reinforced by relatively brittle fibers which have high stiffness and strength. A list of commercially available filaments and description of the manufacturing techniques and their properties have been given by Kreider (1974). Both continuous, aligned fibers and short, chopped fibers have been used to achieve high strength and stiffness. The reinforcement mechanisms of the short fibers are analogous to those which are operative for particles and relatively isotropic properties can be achieved.

Continuous, aligned fibers have been used to form laminated sheets which have directional mechanical properties. A variety of techniques have been employed to produce these laminates. These include infiltration of molten metal into a fiber network and use of preforms of tapes which can be used to form laminates. Repeated layers can be fabricated to form plates by the use of diffusion bonding. Filament winding procedures can also be employed to manufacture axisymmetric bodies.

The type of fiber selected for a given metallic composite depends upon the type of service conditions which are to be encountered. The elevated temperature properties of a titanium alloy, for example, can be enhanced by the use of reinforcing filaments (Metcalf, 1974).

Schijve *et al.* (1979) have described the fabrication of laminates which consist of a number of aluminum alloy sheets which are adhesively bonded together. Schijve (1993) also has described laminates for which aluminum alloy sheets are reinforced with glass or ARAMID fiber – adhesive prepregs. In some service components the stress state developed is primarily uniaxial. In such applications components with unidirectional fiber layups can provide effective substitutes for metals which have been used in the past. Use of metal composite plates can also, however, be used in applications in which biaxial stress states are developed. Requirements for this type of loading condition can be met by the use of metal composites in which successive laminae with different filament orientations are introduced (Schijve, 1993).

One common requirement for manufacturing metallic composites which can be used with confidence in engineering applications is the development of fabrication procedures which eliminate voids and porosity. These can serve as flaws from which failure can begin. This has been a primary requirement in all of the developmental programs which have been initiated.

As the technology of metallic matrix composite fabrication evolved, it became clear that the properties of the interfaces between the reinforcing components and the metal matrices could significantly affect the mechanical properties. For fiber composites the strength of the interface bond determines the conditions for debonding. After debonding, frictional sliding promoted by micro-residual compressive stresses across the interface can affect the stress gradients in both the matrix and the fiber. The consequences of fabrication procedures must, therefore, be understood in order to control the chemistry and microstructural features of transitional phases between metal matrices and the reinforcing fibers (Fishman, 1991).

The compatibilities and relative ductilities of particulate reinforced metal composites also must be carefully examined. Here again, micro-residual stress gradients can be expected to develop. Also, the de-cohesion of an interface bond or the brittle fracture of a particulate can serve as sites from which a failure mechanism can be activated.

## 3.5.2 Fatigue

The increases in stiffness and strength which can be achieved by use of reinforcing elements in metal matrices are often cited as reasons for considering them for use in structural components. The primary emphasis here, however, is on the mechanisms which are operative in fatigue and crack extension for this class of solids.

A collection of articles which provide a comprehensive examination of the properties of metal matrix composites can be found in the proceedings of a symposium edited by Dvorak (1990). Interface properties, damage mechanisms, inelastic behavior and fracture are covered in these articles. A comprehensive review of experimental and analytical studies of fracture in metal matrix composites has been presented by Ochiai (1989).

Kelly and Bomford (1969) presented a theory for determining stress versus cycles to failure curves and applied it to composites with a copper matrix reinforced by continuous tungsten wires. The matrix was assumed to fatigue in the same manner as unreinforced copper. Their theory predicted that fatigue strength increases with increasing volume fraction of fibers. The lives which they predicted were, however, shorter than those found experimentally. They attributed the difference to the fact that they used an elastic – perfectly plastic matrix material model instead of an elastic – strain hardening model. Their research did not explicitly address the crack propagation phase of fatigue life.

Schijve *et al.* (1979) have reported the fatigue crack growth behavior of laminates which consist of a number of sheets of the aluminum alloy 20224-T3 bonded together. They found that the life with a central through crack was about 60% longer for the laminated material than for a solid sheet of the same thickness. They attributed the difference to the fact that the individual laminae were in states of plane stress for which a crack grows more slowly than in the plane strain state which was developed in the solid plate. They also compared the growth of surface cracks in a laminate with that for a solid sheet. A surface crack in a solid sheet was found to grow much more rapidly than in a laminated sheet. In the solid sheet continuous crack penetration and extension occurred. In the composite the crack growth extended in the first laminate sheet only, and then formed delamination branches in the adhesive. Thus the adhesive layers form barriers to crack extension. Additional cracks eventually formed in interior sheets, but were not extensions of the initial crack.

Schijve (1993) has also described the fatigue properties of the laminates which consist of alternate layers of aluminum alloy sheets and glass or ARAMID fiber – adhesive prepregs. Both unidirectional fiber layups and cross-ply [0/90] layups are made and are commercially available. The cross-ply layups are intended for applications in which tensile, biaxial stresses are encountered.

Fatigue crack tests have been conducted in unidirectionally reinforced metal – fiber laminates. The loading was applied in the direction of fiber layup and cracks were normal to the loading direction. It was found (Schijve, 1993) that cracks in alternate metal layers were bridged by adjacent fiber layers. The bridging action was credited with reducing the stress intensity factor at the crack tips of the individual metal layers.

The failure mechanisms which develop between metal matrices and reinforcing fibers have been identified and modeled for unidirectional, continuous fibers. For loading normal to the direction of the fibers a strong interface bond is required for good transverse mechanical properties. A strong bond does not, however, promote good longitudinal strength. The properties responsible for this behavior include the relative stiffnesses and strengths of the matrix and the fiber, the interfacial bond strength, and the residual stress state which affects the frictional force between the matrix and the fiber during fiber pullout.

Experimental and analytical results have generally indicated that good resistance to crack extension is achieved when partial debonding occurs at the fiber matrix interface behind an advancing crack tip (Hutchinson, Mear and Rice, 1987; Cox and Marshall, 1990; Hutchinson, 1990; Gupta *et al* ., 1992). The mechanisms which are operative are depicted in Fig. 3.7. If a crack advances through the matrix without debonding, the stresses in the fibers will become very high as the crack begins to open. If fibers fracture, the benefits of their reinforcing action are lost. If partial debonding occurs, as shown in Fig. 3.7, pullout can occur and the stresses in the fibers can be below the fracture stress. The benefits of bridging can then be realized. This introduces tensile crack face forces, and the local stress intensity factor which ensues must be subtracted from the global stress intensity factor produced by the external loading. The crack face forces introduced by the fibers depend upon the lengths of the interfacial debonds and the interfacial friction stresses. Ultimately, fibers well behind the crack tip fracture as the crack faces spread apart. This action is indicated by the last two fibers on the left. The features of bridging have been verified by experimental studies (Sensmeier and Wright, 1990; Barney, Cardona and Bowen 1993; Davidson, 1993). Models designed to describe the effect of bridging on the crack tip stress intensity factor in brittle matrix composites have been developed by Budiansky, Hutchinson and Evans (1986), McCartney (1987), and Marshall and Cox (1987).

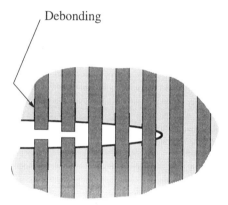

**Fig. 3.7** The fiber bridging mechanism

As a crack front propagates across a unidirectional metal matrix composite under tensile load, it must go around fibers. In a multi-layer layup it follows that the crack front is not ideally straight as might be inferred from the two-dimensional representations which are commonly used to describe fracture behavior. In particulate and whisker reinforcements an advancing crack can go over, under, around or through individual particles or whiskers. An advancing crack plane will then be wavy rather than planar and the crack front will not be straight. These deviations from the idealized geometry can be expected to result in an increase in the resistance to crack advance. Analyses which examine the consequences of deviations have been performed by Rice (1985) and by Bower and Ortiz (1990).

Shang, Yu and Ritchie (1988) have studied the roles of these mechanisms in a powder metallurgy aluminum alloy (similar to 7091) which was reinforced by 20% volume fraction additions of either fine or coarse silicon carbide particles. They found that the operative mechanisms depend on the particle size and the levels of cyclic loading. At low load levels crack surface roughness was greater for the coarse particulate, and the resulting obstruction to closure reduced the effective range of loading and produced smaller crack growth rates. Although some bridging was observed in the form of matrix ligaments around broken particles, the general conclusion was that the fatigue crack growth properties of this particle-reinforced composite was comparable but not superior to the corresponding unreinforced alloy.

Kumai, King and Knott (1990) investigated the fatigue crack growth of 2014 aluminum alloy composites with SiC particulate reinforcement. They found that dispersed SiC particles acted as crack arresters for small cracks. The rate of growth of long cracks, however, was found to be greater than that observed for the monolithic alloy. This composite also had a low fracture toughness value which was attributed to a weak bonding of the SiC/matrix interface.

Short fiber or whisker reinforcement is particularly attractive as a means of extending to higher temperatures the use of alloys which soften as the exposure temperature increases. Improvements in resistance to both creep and fatigue at elevated temperatures are the goals of some of the efforts in the development of the technology of whisker reinforcement.

The microstructure developed for short fiber composites is dependent on the fabrication procedure used. Extrusion, for example, promotes a directional orientation of the fibers (Greenfield, Fang and Orthlieb, 1991). The resistance to fatigue crack growth in composites is best when loading is applied in the extrusion direction. Cracks propagating normal to the thickness direction encounter crack tip shielding and crack deflection is promoted.

As with particulate reinforcement, interactions between whiskers and between fibers and the matrix influence the local states as well as the global properties. Tvergaard (1990) has presented a finite element analysis of the deformation and debondiing in metal matrices reinforced by short fibers. As with particulate reinforcement, the non-flat crack surface may, for whisker reinforcement, require an application of the type of analysis performed by Rice (1985).

Finally, metal matrix composites loaded in compression can exhibit a response mechanism which is not observed under tensile loading (Huang and Wang, 1988). If upper and lower segments of a composite develop opposite, horizontal displacements, as shown in Fig. 3.8, localized kinking can occur. Analyses designed to exhibit this behavior have been proposed by Budiansky and Fleck (1993) and Slaughter and Fleck (1993).

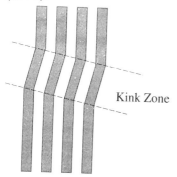

Kink Zone

**Fig. 3.8** Localized kinking during compressive loading

## 3.6 POLYMERIC MATRIX COMPOSITES

Polymeric matrix composites with fiber reinforcement began to be manufactured during the 1940 s. Applications in structural components began to increase rapidly after 1970. Although the production costs tend to be greater than the materials – usually metals – for which they are substitutes, their light weight, stiffness and strength make them very attractive. These composites can, for example, have strength to density ratio values which are four to six times greater than steels and aluminum alloys. Also, they have stiffness to density ratio values which can be three to five times greater than steels and aluminum alloys. The primary limitations on the use of polymeric matrix composites are associated with the relatively high production costs and the need to develop essentially new design concepts. The fabrication procedures and the types of failure mechanisms encountered with composites have required the development of both new production processes and new design methods. Agarwal and Broutman (1990) have provided a comprehensive treatment in which the fabrication, properties and analysis of polymeric matrix composites are covered.

A number of references on the design and the mechanical behavior of composite materials are available. These include books by Ashton, Halpin and Petit (1969), Jones (1975), Johnson (1979), Carlsson and Pipes (1987) and Carlsson (1990).

### 3.6.1 Fabrication

Polymeric matrices are usually classified as being either thermosetting or thermoplastic. Thermosetting polymers are processed to have crosslinking between molecular chains. Thermoplastic polymers generally do not have crosslinking. Epoxy and polyester resins are used to form thermosetting matrices. Epoxy resins are most commonly used for applications which require improved mechanical properties. The limiting exposure temperature for epoxy resins is about 200 °C .For service temperatures of 315 to 370 °C polyimide and polybenzimidazole resins can be used.

Thermoplastic resins range from common industrial plastics, such as polyethylene, nylon, polystyrene, polyester, polycarbonate and acrylic, to resins with better mechanical properties, such as the thermoplastic polyimides. These latter materials have greater impact resistance than the thermosetting materials. Good summaries of the molecular structure and processing details for both the thermosetting and thermoplastic materials are given by Strong (1989) and Sheldon (1982). Thermoplastic resins and their use as matrices in composites are comprehensively covered by Béland (1990).

Often, the reinforcing elements in polymeric matrix composites are continuous filaments or fibers. The most commonly used glass fiber reinforcements are E-glass, S-glass, C-glass and quartz. Compounds described as coupling agents can be applied to the surface of the fibers to provide improved coherency between the fibers and the matrix. The application of these agents can result in significant improvements in strength.

Carbon and graphite fibers are also available and they can be used when large increases in the values of stiffness are desired. The ratio of the modulus of carbon fibers to that of glass fibers can be as large as 4.7. Both glass fibers and carbon fibers have a tensile strength which can exceed those of ultra-high-strength steels by a factor of about two. Surface treatments which clean and add chemical interphases to carbon fiber surfaces can be utilized to enhance the interfacial bond strength between the fibers and the matrix. Organic coatings or sizes can also be applied to improve interfacial strength (Sheldon, 1982). The strength of an interface determines the mechanics of load transfer between the fiber and the matrix. Composites with a weak interface tend to have low strength and stiffness, but high resistance to fracture. Strong interfaces produce composites with high strength, but they tend to be brittle. A good introduction to the structure and properties of fiber-matrix interfaces is presented in chapter 3 of Hull (1981).

Organic fibers based on ARAMID technology have also been used as reinforcing elements in composites. Kevlar, a DuPont fiber, has been widely used in commercial products. By extensional processing, Kevlar can achieve a stiffness which is about two-thirds that of steel. Its strength to density ratio is substantially greater than that for steel.

Polymeric matrix composites are fabricated in a variety of forms. The processing temperatures are lower than those for metal and ceramic matrix composites, so the technological problems involved introduce less restrictions on the types of processes which can be utilized. The processing temperatures can, nevertheless, introduce problems. When two intimately bonded materials with different thermal coefficients of expansion are exposed to a change in temperature, stresses are introduced. The bimetallic strip used in instruments to measure temperature is an example in which this mechanism is exploited. When composites are cooled after curing, differential contraction introduces thermal stresses. There are also, moreover, stresses introduced by resin shrinkage. Thus, the total residual stresses which are introduced are a combination of stresses due to thermal contraction and to shrinkage. Since these stresses are internally developed, they must be internally balanced. There will, therefore, be both tensile and compressive stresses which will vary on a scale which is of the order of the fiber dimensions. The residual microstresses thus generated can sometimes be large enough to cause microcracking. Stresses introduced by externally applied loads must, of course, be superimposed on the residual stresses. A review of some of the analytical methods for estimating the residual stresses is presented in an article by Chamis (1974).

Another form of residual stress can be produced by an absorption of water from a humid atmosphere. The absorption process depends upon the relative humidity, the temperature and the diffusion coefficient of the matrix material. The moisture absorbed leads to a swelling of the matrix material, and it is affected by the fiber geometry and the volume fraction of the fiber/matrix mixture. In thermosetting polymers the effects are also dependent on cross-linking chemistry. Since a uniform distribution of the absorbed moisture is not likely, stress gradients which result in residual stresses can be developed.

In addition to the introduction of residual stresses, reductions in the values of the elastic properties of the matrix material can occur. These reductions can adversely affect composite properties for which matrix stiffness contributes to strength, e.g. compressive strength and interlaminar shear properties.

For more detailed presentations of issues related to environmental effects the reader may refer to proceedings of a conference edited by Vinson (1978).

The preparation of fibers for use as reinforcing elements in composites borrows technology developed in the textile industry. The reinforcement elements can be used as individually separated fibers, untwisted bundles of fibers (tows), twisted bundles (yarn) or woven fabrics. Braiding can also be used to interlace yarns or tows in tubular shapes.

The fabrication of composites can be facilitated by the preparation of mats of fibers in a binder or tapes of unidirectional fibers imbedded in a resin. The integration of the reinforcing elements into a polymeric matrix is achieved by a variety of techniques. The manufacturing of composite components for structural applications can be accomplished by successively applying layers of preimpregnated tapes, mats or woven fabrics to form sheets, plates or shells. The use of prepreg tapes introduces the possibility of changing fiber orientation in successive layers and eliminates the directional stiffness and strength properties of unidirectional reinforcement.

Axisymmetric shells can be fabricated by a process in which bundles of filaments are wound onto a rotating mandrel. The carriage supplying the filaments moves back and forth, parallel to the axis of the mandrel, and lays down the filaments in predetermined angles. The filaments can be either impregnated as they are applied or the bundles can be preimpregnated.

Composite materials have been fabricated for use in a variety of configurations where they are adapted to fulfill geometric and loading requirements. One good example which illustrates the potential of this type of composite is in sandwich construction. Considerable experience has been acquired from the use of metal sandwich construction. The advantage in the use of composite sandwich construction derives from a realization of gains in bending stiffness and strength to density ratios. Failure mechanisms in sandwich construction include consideration of the effects of tensile and compressive in-plane loading of the faces and the integrity of the bond between the faces and the core. The mechanisms can include delamination and wrinkling of the faces.

A variety of other fabrication procedures have been developed. These include several processes which are capable of providing high production rates. The pultrusion process involves drawing impregnated, continuous fibers through a die. In matched die molding a reinforced mat is compressed between male and female dies. Composite products can also be produced by injecting particles or whiskers and a resin into a mold. This is described as an injection molding process. Descriptions of these and other manufacturing processes are described by Strong (1989), Béland (1990) and Agarwal and Broutman (1990). Also, a concise summary of manufacturing processes for these composites is presented in table 1.5 of a text by Hull (1981).

### 3.6.2 Fatigue

The fundamental differences between metals and polymeric solids has been alluded to previously. The differences between metals and the polymeric composites, which are viewed as substitutes for metals, are substantial. Although the primary concern here is about fatigue, it is enlightening to compare briefly the differences in the responses of these two types of materials to monotonic as well as cyclic loading.

Monotonic loading of ductile, structural metals initially involves a linear elastic response for which tensile and compressive loading are essentially the same and are reversible, i.e. no permanent changes in size or shape are introduced. Beyond the proportional limit, the irreversible, inelastic deformation which occurs under both tensile and compressive loading is a manifestation of crystallographic slip. Only for relatively large strains do the response mechanisms for tension and compression change, i.e. ultimate fracture for tension and crushing or barreling for compression. (We exclude here buckling in metals, which depends upon column geometry as well as material properties.)

The mechanisms developed for polymeric composites are quite different from those for the metals for which they are being considered as substitutes. Evidence indicates that degradation in the form of localized microcracking can begin at relatively low loads. The damage incurred depends upon the relative mechanical properties of the fibers and the matrix, the interfacial bond strength between the fibers and matrix, and the interlaminar strength. Ultimately, progressive delamination can occur. The form of this mechanism differs for tensile and compressive loading. The delamination which occurs can be a form of stable crack growth and it can progress under mode I, mode II or mixed mode I and II conditions. A good review of interlaminar mode I fracture testing has been presented by Davies and Benzeggagh (1989). A comprehensive description of the performance and analysis of interlaminar mode II experiments has been given by Carlsson and Gillespie (1989).

Cyclic loading in metals results in a crack initiation phase followed by a stable crack growth phase. In polymeric composites localized microcracking leads primarily to progressive delamination. The plane of the crack extension in metals is normal to the direction of tensile loading. In composites, however, the planes of the growing delaminations are parallel to the axis of loading. Also, in metals, the maximum load during cycling must be tensile for crack growth. Both tensile and compressive cyclic loading in composites can lead to progressive delamination. The failure mechanism for compressive loading in composites can involve both localized, micro-buckling of fibers and global buckling which can lead to progressive delamination. Also, in unidirectional composites compressive loading can lead to a localized kinking which extends through the cross-section. For brittle carbon fibers this can result in fiber fractures at each end of the kink region. For less brittle Kevlar fibers, however, kinking without fracture can occur (Hull, 1981).

A number of tests are used for determining in-plane shear properties. These include torsion tests on filament wound tubes (Pagano and Whitney, 1970), off-axis tests on unidirectional composites (Chamis and Sinclair, 1977), and tensile tests on $[\pm 45]_s$ coupons (Rosen, 1972). The Iosipescu shear test, which produces in-plane shear across a doubly edge notched specimen loaded in bending, is also used (Adams and Walrath, 1987; Pindera *et al.*, 1987). The objective of these tests is to determine the in-plane shear stiffness and strength.

To examine failure mechanisms for combined loading conditions, Hinkley and O'Brien (1991) conducted tests on plates of a quasi-isotropic graphite epoxy laminate subjected to tensile and torsional loading. They found that the addition of a torsional load reduced the value of the tensile load at which delamination failure occurred. For pure tensile loading the primary matrix crack which led to delamination was normal to the loading direction. For torsion or combined torsion and tension, however, the primary matrix crack was observed to be inclined to the tensile loading direction.

For design goals in which lifetimes of the order of $10^6$ to $10^8$ cycles of loading are anticipated aluminum alloys have endurance limits which are substantially less than their static strengths. Fiber reinforced composites have, by contrast, exhibited the ability to withstand such large numbers of cycles without being subject to large reductions in load carrying capacity. This property has created considerable interest in the substitution of composites for some aluminum alloy components in aerospace applications.

The consequences of exposure to cyclic loading on a composite can be viewed at both microscopic and macroscopic levels. Ultimately, a designer is, however, primarily interested in the residual strength which exists after damage has been incurred.

Gao, Reifsnider and Carman (1992) have developed micromechanical models in which statistical distributions of fiber and matrix strength and interfacial bonding strength are assumed. Criteria for different failure modes are then adopted to evaluate the degree of property degradation at each level of loading. The states resulting from stress redistributions may be examined for both monotonic and cyclic loading. Gao (1992) has suggested that the models can be used to conduct parametric studies and sensitivity analyses which may provide guides for the optimum design of composites.

Hertzberg and Manson (1980, 1986) have summarized fatigue test results for polymeric composites reinforced with particles and with whiskers. They cite data of Mandell *et al.,* (1983) in which it is demonstrated that additions of glass and carbon whiskers to thermoplastic, injection molded polysulfon provide substantial improvements in the stress versus cycles to failure properties. Also, as with the metallic matrix composites discussed in section 3.5.1, the resistance to fatigue crack propagation is increased by crack tip shielding.

Often, biaxial stress states are encountered, and then the properties transverse to the direction of the lay-up for unidirectional composites can be expected to be inadequate. Hashin and Rotem (1973) conducted stress versus cycles to failure tests on epoxy matrix composites with continuous, unidirectional glass fibers to examine the effect of off-angle tensile loading. For tests in which the tensile direction was oriented at an angle of 60 degrees to the fiber direction, the stress values were less than ten percent of those for tests which were tensioned in the fiber direction.

The limitations of unidirectional fiber layup can be eliminated by the use of cross-ply layups, and numerous experimental and analytical studies have been performed on such laminates. Although the use of cross-ply layups provides an improved response to loading conditions which introduce biaxial stress states, they can lead to failure modes which are not present in unidirectional layups. Differences in stiffness and strength between adjacent cross-ply laminates introduce response mismatches which ultimately can result in matrix ply cracking and delamination (O'Brien, 1991). This behavior is discussed in more detail in Chapter 7.

One measure of degradation or damage in cross-ply laminates is residual strength. Another measure has been obtained by continuously monitoring stiffness during cycling (Reifsnider, Schulte and Duke, 1983). Talreja (1986) and Hashin (1986) have developed analyses in which crack densities and delaminations are introduced to model observed reductions in stiffness. Leong and King (1991) have shown that damage accumulation in the form of fiber – matrix debonding and transverse cracking during cyclic loading can be correlated with stiffness reduction in $[0,90]_{2s}$ fiber reinforced epoxy composites. Chen, Harris and Reiter (1991) conducted fatigue tests on cross-ply and unidirectional laminates. They found that the damage incurred, as reflected by residual stiffness and strength, was greater for the cross-ply laminates. For biaxial loading the apparent superior resistance to damage of the unidirectional laminate would, of course, be lost.

The damage mechanisms which develop under cyclic tension differ from those for compressive loading. Schultz and Reifsnider (1984) have conducted fatigue tests which included combinations of tensile – tensile, tensile – compressive and compressive – compressive excursions. Graphite epoxy laminates with six groups of [0/45/90/–45] stacking sequence were evaluated. They found that the damage incurred for tension – compression combinations was more severe than would be predicted by superposition of damage from tensile cycling alone and compressive cycling alone; i.e., damage inflicted during tensile excursions, for example, was exacerbated by the compressive excursions. Matrix cracking and interlaminar cracking which led to delamination were observed to occur preferentially near the specimen edges and near the top and bottom surfaces. Localized buckling during compressive excursions was cited as the mechanism which caused an increase in the rate of delamination growth.

Adam, *et al.,* (1991) have also compared results for fatigue tests with tension – tension, compression – compression, tension – compression loading. For a laminate with a crossply layup of carbon fibers with a bismaleimide matrix, they found that the damage incurred, as measured by fatigue strength, was greatest for $R < 0$ loading.

As damage in the form of matrix and fiber cracking occurs, there is a redistribution in local stresses which can lead, even under monotonic tensile loading, to the development of local mixed mode states. This ensues, for example, when a transverse crack is blocked at a lamina, and turns to initiate interlaminar cracking, which in turn evolves to form a delamination. This type of mechanism initiates at edges and from notches and holes (O'Brien, 1982; Trewetthey, Gillespie and Carlsson, 1988; Sbaizero *et al.*, 1990). The delamination crack tip has a stress state which is locally characterized by both $K_I$ and $K_{II}$ stress intensity factors. In addition to shear which is developed between laminates, there are forces normal to the crack faces which result from bridging of fibers and matrix ligaments (Sbaizero *et al.*, 1990; Bordia *et al.*, 1991).

**Fig. 3.9** Fiber bridging in a split log

A form of delamination bridging which may be readily recognized by anyone who has split a wooden log is shown in Fig. 3.9. The fibers which bridge the fracture surfaces perform the same function as fibers which bridge a delamination gap in laminates. That is, they provide a resistance to continuing crack extension.

Delamination in composites can include all three fracture modes, i.e. opening, sliding shear and scissoring shear. Measurements of interlaminar toughness have been made by use of a double cantelever beam test (Whitney, Browning and Hoogstederr, 1982) in which delamination starts at an insert which represents an initial delamination crack. Although the initial crack extension may be governed by a mode I state, toughness has been found to increase with continuing delamination (O'Brien 1991). Bao, Fan and Evans (1992) have conducted a finite element analysis of the delamination process, and their results indicate that continuing extension occurs under mixed mode I and mode II conditions. They attribute the observed behavior to the development of tractions on the delamination crack faces, and they suggest that the tractions are developed by bridging fibers and matrix ligaments.

Mode II delamination toughness has been measured on an end notched flexure test (Kageyama *et al.*, 1991). A test procedure in which a mixed mode I and mode II state is developed has been introduced by Reeder and Crews (1991). By changing the location of the loading point on the fixture, the ratio of mode I and mode II can be varied.

The objective of the delamination tests is to isolate the behavior from other damage mechanisms and to obtain, thereby, data which characterize the delamination process. The same tests used to determine delamination toughness under monotonic loading have been used to evaluate resistance to delamination under cyclic loading. The complexities associated with retardation effects due to bridging and the presence of mixed mode states, however, introduce uncertainties in the interpretation of the data. Also, plots of crack growth rate versus energy release rate, $G$, on log – log plots are very steep (Martin and Murri, 1990). Small changes in $G$ can result in very large changes in crack growth rate. This indicates that the composites which they tested were not damage tolerant to delaminations.

It has been noted that interlaminar cracking which can precede delamination often occurs at free edges. An analysis which provides insight into this behavior has been provided by Pipes and Pagano (1970). Whereas elementary lamination theory considers only in-plane stresses in each lamina, Pipes and Pagano performed an analysis in which through-the-thickness, normal and shear stresses were included. This more comprehensive treatment revealed that an interlaminar shear stress acting parallel to the free edge is developed for a tensioned, angle-ply laminate. They found that the magnitude of the shear stress at the edge varied with fiber orientation. An examination of test data on angle-ply laminates has indicated that the shear stresses predicted by Pipes and Pagano can be invoked to explain the onset of interlaminar cracking adjacent to the free edges (Rotem and Hashin, 1975). It should also be noted that since these interlaminar shear stresses extend into the laminate a distance of the order of the laminate thickness, there is a specimen width effect, i.e. the strength has been observed to increase as the thickness to width ratio decreases (Hull, 1981).

A recognition of the vulnerability of angle-ply laminates to edge delamination has led to efforts to develop fabrication techniques which can improve the resistance of the edges to this form of damage. Jones (1991) has summarized results for a number of proposed delamination-suppression design concepts (Mignery, Tan and Sun, 1985; Chan, Rogers and Aker, 1986; Chan and Ochoa, 1988; Howard, Gossard and Jones, 1989; Sun and Chu, 1991). These include edge reinforcement and edge modification. Edge reinforcement can be achieved by use of an edge cap, trans-laminate edge stitching and interleaved adhesive layers. Edge modification consists of reducing the edge thickness by internal termination of a mid-thickness ply, or by introducing edge notches. On the basis of test results, Jones concludes that edge caps, interleaved adhesive layers and ply termination can delay delamination and increase both the static strength and the fatigue life. Edge notching suppresses delamination, but both increases and decreases in static strength have been observed for four different laminates. Fatigue data are not available.

The effect of environment on polymeric matrix composites was alluded to in section 3.6.1. It was noted that absorption of water from a humid atmosphere could lead to a swelling of matrix material which in turn could introduce residual stresses. The sensitivity to environment can also have an adverse effect on the resistance to delamination during long term cyclic loading. Marom (1989) has presented a review of the effects of environment on delamination resistance.

The stiffness and strength to density ratios of reinforced polymeric composites clearly make them attractive choices for use as structural components. Aerospace applications with their emphasis on weight saving requirements are particularly appealing. Improvements in the resistance to interlaminar edge cracking and delamination are the objectives of continuing research and development. It should also be recognized that these efforts include not only the composite fabrication techniques, but also the details of incorporating the composite components into structural systems. These latter details involve cutting, drilling, machining and joining operations (Strong 1989, Béland 1990).

Results obtained from test coupons cannot generally be used to predict the detailed behavior of manufactured, built-up structures. In aircraft construction, for example, built-up panels subjected to compressive loading and to shear loading are common. For both of these cases the panels are designed to allow for buckling behavior which is not considered to be a failure (Timoshenko and Gere, 1961). For the compressed panel, increases in load beyond buckling are absorbed by regions adjacent to the supported edges which are parallel to the direction of loading. For shear loading a diagonal tension is developed and it extends the load carrying capability of the panel. Under the variable loading encountered in aircraft, these panels are subjected to repeated postbuckling which may induce fatigue failure.

The problem of fatigue which is induced by repeated postbuckling in composite construction has been investigated by Weller and Singer (1990). They conducted extensive experimental studies of the durability of a variety of graphite/epoxy panels subjected to compressive loading and to shear loading. Although repeated postbuckling was observed to introduce delamination, catastrophic failure did not ensue. Static tests on damaged panels indicated that the residual strength was reduced by 20% or less for the shear panels. No reduction in residual strength was observed for the compressed panels. Also, comparisons of data from other repeated buckling tests indicated that composite panels were less sensitive to fatigue damage than comparable metal panels. From their results Weller and Singer (1990) concluded that the use of composite panel construction could lead to improvements in structural efficiency.

Methods of designing composite components based on the damage tolerant philosophy which is widely used in metal structures have been proposed. These begin by assuming that matrix cracks are present initially, and then proceed to determine the tolerance of a component to the crack damage (O'Brien 1990). This philosophy of design is considered in Chapter 7.

## 3.7 CERAMIC MATRIX COMPOSITES

It is observed in the section on brittle solids that ceramics are considered to have some advantages for use in elevated temperature environments. The primary requirement in addition to a resistance to environmental attack is that such ceramic components possess both sufficient strength and sufficient resistance to time dependent inelastic flow or creep. Cyclic loading at elevated temperatures can also be expected to lead to a coupled creep – fatigue behavior. For the current state of progress, however, an understanding of the interaction of these mechanisms is primarily qualitative in character.

### 3.7.1 Fabrication

In spite of improvements in the properties of the ceramics which are being produced, there is considerable interest in achieving improvements through the development of ceramic matrix composites. Ceramic composites which are reinforced by continuous fibers, whiskers or particles are being produced. A variety of fabrication procedures have been developed. These include hot pressed and sintered mixtures of oxide powders which are reinforced by either particles or whiskers.

Laminated ceramic composites have also been fabricated by tape casting techniques which use whiskers for reinforcement. Lamellar composites are then produced by stacking layers with the compositions and orientations which provide the desired mechanical and physical properties. Densification is performed by hot pressing. This technique is not limited to the production of flat plates, since the flexible tapes can be formed on mandrels to fabricate complex shapes (Amateau, Conway and Bhagat, 1993).

Unidirectional and cross-ply composites of continuous silicon carbide fibers in a $Si_3N_4$ matrix have been fabricated by use of a prepreg tape layup procedure. The composite preforms are consolidated by hot-pressing (Yang *et al.*, 1991). A comprehensive review of the fracture behavior of ceramic matrix composites with continuous fiber reinforcement has been presented by Davidge (1989). An extensive review of whisker reinforced ceramics appears in an article by Warren and Sarin (1989).

Kochendörfer (1993) has described a process in which liquid silicon is infiltrated into a porous carbon – carbon skeleton to produce a composite of carbon fibers in a silicon carbide matrix. Woven fabrics and braided tubes are respectively used to produce plates and nozzles.

### 3.7.2 Fatigue

As noted previously, an objective of the developments of ceramic matrix composites is to produce materials for components which are exposed to elevated temperature service conditions, *i.e.* temperatures between 1000 °C and 1500 °C. As a consequence, the emphasis is on the development of useful strengths and creep resistance at these temperatures. Improvements in these properties over those for unreinforced ceramics are dependent on the effectiveness of crack tip shielding mechanisms of the type referred to in the discussions of metal matrix composites. The mechanisms which may be operative include crack bridging with debonding and fiber pull-out and crack deflection (Campbell *et al.*, 1990; Rodel, Fuller and Lawn, 1991).

A comprehensive review of the fracture behavior of ceramic matrix composites with continuous fiber reinforcement has been presented by Davidge (1989). An extensive review of whisker reinforced ceramics appears in an article by Warren and Sarin (1989).

Acquisition of fatigue data on ceramic matrix composites is limited. Holmes (1991) has conducted cyclic uniaxial tensile tests on $Si_3N_4$ which was unidirectionally reinforced by $SiC$ fibers in the temperature range 1000 to 1200 °C. When the maximum stress was greater than the proportional limit, there was a progressive deterioration in effective stiffness. The ultimate failures involved a coupled creep and fatigue behavior.

Ogawa, Tokunori and Keiro (1993) have conducted fatigue crack growth tests on alumina reinforced by $SiC$ whiskers at room temperature. They fitted their data to a power law of the Paris type and obtained an exponent of 80.

Han and Suresh (1989) and Suresh (1990) also conducted fatigue crack tests at 1400 °C on alumina reinforced by $SiC$ whiskers. They obtained Paris exponents of 7 and 4 respectively. Since results for static tests and two different cycle frequency tests indicated that creep mechanisms were operative at the test temperature, there is a question about the soundness of using a Paris plot to correlate the data.

From the large values of the Paris exponents it may tentatively be concluded that the crack growth behavior of the reinforced ceramics is similar to that of the unreinforced ceramics, *i.e.*, very small changes in load can result in large changes in crack growth rate. It also may be surmised that for elevated temperature applications, the details of the creep – fatigue coupling behavior need to be considered in evaluations of test data.

The statistical nature of the brittle fracture problem has been discussed in a previous section. The initial flaw distributions in brittle solids form the basis for determining the probabilities of failure according to the statistical theory of Weibull. In the discussion of cyclic (thermal) loading of brittle solids it was observed that cyclic loading below critical loading values could be expected to amend the flaw distribution, *i.e.* flaws in the form of microcracks could develop during loading and they would be added to flaws which were present initially. In an unreinforced brittle solid, fracture is assumed to occur at a critical flaw and then propagate catastrophically. That is, it is tacitly assumed that the initial flaw distribution is not modified by cracking prior to failure. A solid classified as brittle cannot tolerate an activated crack.

The above scenario is not, however, applicable to reinforced brittle solids such as ceramic composites. During monotonic loading and prior to final fracture, progressive, non-critical microcracking can occur. The behavior implied has been examined in a probabilistic modeling of damage in ceramic matrix composites by Lamon (1992). He chose as a model a microcomposite in which a single fiber embedded in a ceramic matrix was subjected to tensile loading. An analysis of a microcomposite with a silicon carbide matrix and a silicon carbide fiber with a carbon interface was performed to examine the relationship between interfacial properties and mechanical behavior. The results of the probabilistic modeling indicated that weak interfaces limit matrix cracking and strong interfaces tend to result in fiber fracture. The modeled behavior is in agreement with that which has been observed in ceramic matrix composites. Wagner (1989) has presented a thorough review of statistical phenomena in fibers and in brittle composites. Both experimental and analytical results are discussed.

The implications with regard to cyclic loading or fatigue appear to be relevant to the extent that both monotonic and cyclic loading in ceramic composites involve progressive degradation and the sequence of the events has a statistical character. The elementary model formulated by Lamon can not, however, account for possible changes in flaw distribution resulting from the interaction of events which occur during fatigue in composites with many fibers.

# 4

# Engineering characterizations of safe life

## 4.1 INTRODUCTION

Although many studies on fatigue have been undertaken, the basic methods used in engineering design have their foundation in the pioneering work conducted in Germany and England during the latter part of the nineteenth century. These efforts were alluded to in Chapter 1.

For machinery in which the amplitude of the cyclic loading is more or less uniformly repeatable and the desired lifetimes involve millions of cycles, design can be based on data obtained from tests in which loading is of the type depicted in Fig. 2.3. One of the topics included in this chapter is the stress-life mode of testing which makes use of the loading history shown in Fig. 2.3. A second topic concerns high loading levels and smaller cycles to failure and it is described as a strain-life approach. This makes use of loading histories of the type shown in Fig. 2.4. The methods described evolved as engineering responses to observed fatigue failures in metallic materials. The initial emphasis is, therefore, on applications involving metals. The applicability of available safe life strategies for nonmetallic materials and composites are considered in section 4.5.

## 4.2 THE STRESS-LIFE STRATEGY

### 4.2.1 The stress-life diagram

Uniform, constant amplitude tension – tension (positive stress ratio $R$) and tension – compression (negative $R$) stress-life tests are conducted. The test used most frequently to obtain basic stress-life data, however, is the rotating bend test for which the stress ratio $R = -1$. A cylindrical, hourglass gage section which has a highly polished surface is used. The high rotational speeds which are employed make it possible to accumulate large numbers of cycles in a reasonably short period of time. Tabulations of mechanical properties for commonly used alloys often include fatigue data which are obtained from this type of test (e.g., Osgood, 1982). The stress values used from reversed bending tests are the maximum bending stresses computed from elementary beam theory. For a circular cross-section, therefore, the stress, $\sigma$, is given by the equation

$$\sigma = \frac{4M}{\pi r^3}, \tag{4.1}$$

where $M$ is usually the moment in the center section of a four point bending test, and $r$ is the radius of the cross-section. Note that $\sigma$ is the alternating stress in these tests and that the total stress range is $2\sigma$.

Because of data scatter, multiple tests are conducted at each stress level, and the test data are often summarized on stress – log cycles to failure plots of the types shown in Figs 4.1 and 4.2. It should be noted that these curves correspond to the failure curve shown in Fig. 1.1. Failure is defined as the fracture of a specimen. The crack initiation curve depicted in Fig. 1.1 cannot be experimentally determined.

Curves similar in form to those shown in Figs 4.1 and 4.2 are drawn through median points at each stress level, so they represent, approximately, 50% probabilities of failure.

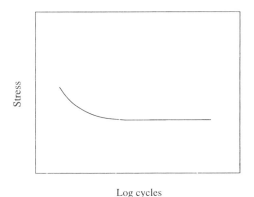

Log cycles

**Fig. 4.1**  Stress versus log of cycles diagram with a fatigue limit

The curves shown in Figs 4.1 and 4.2 represent the two distinct behaviors which are encountered. The curve of Fig. 4.1 decreases asymptotically to a horizontal, constant stress value. Stress amplitudes below this value do not result in failure. Microcracks may develop below this stress level, but growth is arrested. The lower bound stress is described as the endurance or fatigue limit, and these values are often reported in material property tabulations. This limiting stress behavior is most commonly observed for low strength steels.

Most metals exhibit curves of the type shown in Fig. 4.2, i.e. they do not exhibit a lower bound. Most nonferrous metals do not exhibit a lower bound and as a result the endurance or fatigue limits which are reported for them are actually the stress level corresponding to a cyclic life of the order of $10^8$ cycles.

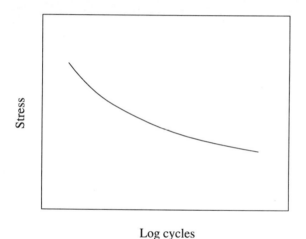

Log cycles

**Fig. 4.2**   Stress versus log of cycles diagram with no fatigue limit

The reversed bending fatigue test provides data for a stress ratio value of $R = -1$. Data for other values of $R$ can be obtained by use of uniaxial tension – tension and tension – compression tests. Values of $R$ can then range from negative values of $R$ to values approaching $+1$. The American Society for Testing and Materials has established standards which are recommended for fatigue testing. These are designated as Standards E466 – E468.

The dependency of smooth bar fatigue behavior on the stress ratio $R$ is important in design and it is considered in section 4.2.3 where it is discussed in terms of the mean value of the maximum and minimum stresses. Other important issues will be considered first, however.

An engineer concerned with fatigue of a component that is being designed must recognize that the testing conditions used to generate the available data probably do not duplicate the operating conditions to be encountered in service. Among the problems that may arise are:

1. The size and geometry of the component may be quite different from that for which data are available. Often, the site of critical concern will involve a stress concentration.
2. The surface condition of the component very likely will not be highly polished. Also, special surface treatments applied to components can result in significant fatigue behavior differences.
3. The manufacturing processes may differ, e.g. test data may have been obtained on drawn bar stock and the component may have been forged. This can result in differences in grain size and texture and the distribution of precipitates and inclusions.
4. The environment for the test specimens may be different from that to which the proposed component will be exposed, e.g. a corrosive atmosphere or an elevated temperature.

A possible solution procedure which accounts for the differences cited is to perform fatigue tests on an actual component under conditions which duplicate those to be encountered in service. Although this approach is sometimes used, it often cannot be economically justified. An alternative is to introduce factors which can be applied to compensate for differences between the test conditions used for data which are available and the actual conditions encountered in service. A familiarity with the effects of the items listed above is essential if this latter procedure is to be acceptable.

The size effect cited in item 1 is analogous to that described in section 3.4 in the discussion of the statistical nature of brittle fracture. For two bars subjected to reversed bending with the same maximum tensile stress, the larger bar has a larger surface area and, hence, more potential sites for the initiation of microcracks. Faupel and Fisher (1981) and Forrest (1962) have cited examples for which endurance limits have been reduced by from 15% to about 50% due to differences in size. Large forgings and castings can be expected to be subject to such size effects. It may also be noted that difficulties in the heat treatment of large components can also lead to variations of the type referred to under item 3 above, i.e. microstructural differences. McClintock (1955) has proposed a criterion for controlling and minimizing scatter in fatigue properties.

The surface condition of a component to be used in a structure or a machine depends on the last operation performed in its fabrication.

These operations may involve machining with a cutting tool, grinding, electro-discharge machining, chem-milling, rolling, drawing, extrusion or forging. Each of the processes produces a characteristic surface. Surface roughness and surface layer residual stresses can result in significant differences in fatigue behavior. Surface roughness can introduce localized stress concentrations which reduce resistance to fatigue crack nucleation. A tensile residual stress on a component surface can adversely affect fatigue crack initiation. A compressive residual stress on the surface can, conversely, delay crack initiation. Juvinall (1967) and Shigley and Mitchell (1983) have discussed the use of factors which can be applied to available data to reflect differences in fatigue limits for different surface conditions.

The recognition that some forms of surface preparation can lead to improvements in the resistance to fatigue crack initiation has led to the development of a variety of methods of surface conditioning. From an engineering perspective, focusing on decreasing vulnerability is analogous to attempts to devise fabrication schemes which reduce the susceptibility of layered composites to the edge delamination mechanism described in section 3.6.2.

Treatments which are used to modify surfaces include mechanical and thermal treatments and electroplating. Mechanical treatments involve cold rolling of the surface and shot peening. Both introduce compressive residual stresses in the surface layer and therefore subtract from the maximum cyclic tensile stress developed in service (Almen and Black, 1963). Shot peening of fillet areas, which are potential sites for crack initiation, is particularly effective in extending fatigue life.

Another example in which compressive residual stresses are introduced occurs in the aircraft industry where tools which expand rivet holes are used. The holes are sites of stress concentration and here also the presence of residual compressive stresses results in reductions of the tensile stresses which are developed by applied loads.

The maximum and minimum stresses in reversed bending are equal and opposite in sign, and the mean stress is zero. If a bar surface has a residual compressive stress, however, the mean stress is nonzero and negative. There is a translation of the stress history by an amount which is equal to the compressive residual stress. The fatigue behavior of two specimens with the same range of stress but different mean stress values is different. Mean stress effects are discussed in section 4.2.3. Thermal treatments in steels include nitriding and carburizing of surface layers. The surface layer has greater strength than the base material and a volumetric increase introduces compressive residual stresses. Both of these result in improved fatigue lives. Induction and flame hardening can also be used to produce greater surface strength and to introduce favorable compressive residual stresses (Juvinall, 1967).

Electroplating steels with either nickel or chrome improves resistance to corrosion, but introduces residual tensile stresses which reduce fatigue life. Shot peening of nickel plated parts can be used to compensate for the loss of fatigue life due to plating alone, i.e. the tensile residual stresses due to plating are removed by shot peening (Almen and Black, 1963).

### 4.2.2 Stress concentration effects

Stress concentration locations, which result from the presence of holes, notches and fillets, can be sites of large, highly localized stresses. These stresses may greatly exceed the nominal stresses produced away from the stress concentration site, and they therefore represent critical locations for the initiation of fatigue cracks. The stress concentration values depend upon the geometric features of the associated discontinuities. For a tensioned plate with an edge notch, for example, the stress concentration depends upon the ratio of the depth of the notch to the radius of the notch. Stress concentrations for a wide variety of cases are contained in texts by Peterson (1953) and by Savin (1961).

The procedure used in the design of components with stress concentrations depends on the level of loading to be encountered. If the amplified stress is small enough to be within the range of high cycle fatigue, a stress-life design procedure can be used. By definition the stress concentration factor, $K_t$, is defined as

$$K_t = \frac{\sigma_{max}}{S}, \tag{4.2}$$

where $S$ is the nominal stress and $\sigma_{max}$ is the stress amplified by the presence of the stress concentration. For a stress-life approach, a fatigue notch factor, $K_f$ is defined as the ratio of the unnotched fatigue strength, $S_e$, to the notched fatigue strength $S_e$ (notched).

Thus,

$$K_f = \frac{S_e}{S_e(\text{notched})}. \tag{4.3}$$

This ratio depends not only upon geometry, but also on the material, so a notch sensitivity factor, $q$, is used. It is defined as

$$q = \frac{K_f - 1}{K_t - 1}. \tag{4.4}$$

$q$ can vary from zero for no notch to one for the case in which $K_f = K_t$. A design value of $S_e$ (notched) can be determined by use of equations (4.2), (4.4) and (4.3). Values of the notch sensitivity factor, $q$, have been suggested by Peterson (1959) and Neuber (1946). Bannantine, Comer and Handrock (1990) have discussed issues involved in the use of the stress-life approach in the design of notched components.

A strain-life analysis of notches based on a procedure proposed by Neuber (1961) can be used for low cycle fatigue loading. This is discussed in section 4.3. A consideration of the growth of cracks which have been initiated at the base of notches is presented in Chapter 5.

### 4.2.3 Mean stress effects

The stress histories for many components do not have a mean stress of zero. Mean values can be either tensile or compressive. The nonzero mean can be due either to externally applied loading or to residual surface layer stresses.

The parameters used to describe mean stress effects are defined in terms of the maximum and minimum cyclic stresses. The mean stress

$$\sigma_m = \frac{1}{2}\left(\sigma_{\max} + \sigma_{\min}\right). \tag{4.5}$$

The alternating stress

$$\sigma_a = \frac{1}{2}\left(\sigma_{\max} - \sigma_{\min}\right). \tag{4.6}$$

These two quantities can be expressed by dimensionless ratios $(\sigma_a/\sigma_e)$ where $\sigma_e$ is the fatigue strength and as $(\sigma_m/\sigma_y)$ where $\sigma_y$ is the yield strength or as $(\sigma_m/\sigma_u)$, where $\sigma_u$ is the ultimate strength. Reference to the values in the denominators provides a basis for developing empirical relations for describing mean stress effects.

A common approach for developing empirical equations is to represent a dependent variable in terms of a power series in an independent variable. To correlate this type of approach with equations which have been widely used we select the ratio $(\sigma_a/\sigma_e)$ as the dependent variable and either $(\sigma_m/\sigma_y)$ or $(\sigma_m/\sigma_u)$ as the independent variable. Thus, we can write, for example,

$$\frac{\sigma_a}{\sigma_e} = C_1 + C_2\left(\frac{\sigma_m}{\sigma_u}\right) + C_3\left(\frac{\sigma_m}{\sigma_u}\right)^2 + \ldots \tag{4.7}$$

If three terms were to be used, three conditions for values of the ratio $\sigma_m/\sigma_u$ would be required to evaluate the coefficients. Only two conditions are usually used, however. One procedure which has been used is to observe that for $(\sigma_m/\sigma_u) = 0$ and $(\sigma_a/\sigma_e) = 1$, it follows that $C_1 = 1$. Then, truncating the series so that $C_i = 0$ for $i \geq 3$, and setting $(\sigma_m/\sigma_u) = 1$ for $(\sigma_a/\sigma_e) = 0$, it follows that $C_2 = -C_1$. Then, equation (4.7) may be written as

$$\frac{\sigma_a}{\sigma_e} = 1 - \frac{\sigma_m}{\sigma_u} \tag{4.8}$$

This equation, proposed by Goodman (1899), has been described as the Goodman equation. The graph of Fig. 4.3 depicts the suggested variation in $\sigma_a/\sigma_e$ as a linear function of $\sigma_m/\sigma_u$. Only the values on the coordinate axes are based on actual experimental values. Thus, when $\sigma_m = 0$, $\sigma_a$ equals the value of fatigue strength for reversed loading. When $\sigma_a = 0$, $\sigma_m$ has as its upper limit the ultimate strength. Soderberg (1939) proposed a linear relation in which $\sigma_u$ in equation (4.8) is replaced by $\sigma_y$.

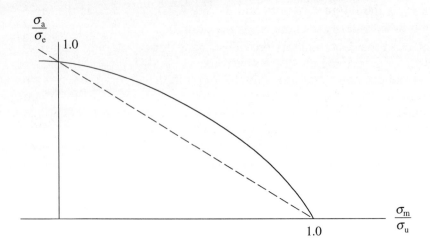

**Fig. 4.3**   The Goodman diagram (dashed line) and the Gerber diagram (solid line)

Since test data plotted on diagrams of the type shown in Fig. 4.3 tend to lie above the straight line, and the end points are considered to be acceptable, it follows that a nonlinear curve lying above that shown might be preferable. By limiting the curve fitting process to the two end point conditions, it follows that an acceptable procedure would be to retain only the $C_1$ and $C_3$ terms. Proceeding as before, it then follows that an equation credited to Gerber (1874) can be written as

$$\frac{\sigma_a}{\sigma_e} + \left(\frac{\sigma_m}{\sigma_u}\right)^2 = 1 \tag{4.9}$$

A plot of this relation is also shown in Fig. 4.3.

If a sufficient number of tests are conducted, plots of the type shown in Fig. 4.4 can be constructed for constant lifetime curves. These are called Haigh diagrams. Master curves which present data in this form are presented for a number of metals in the *Mil Handbook 5*.

The curves of Fig. 4.3 can be extrapolated to the left of the ordinate axis to represent the effects of compressive mean stresses. It is of interest to note that if data were used to establish the slope of the curve at $\sigma_m = 0$, a third condition would become available, and coefficients for the first three terms of the series of equation (4.7) could be evaluated. Equation (4.9) has a slope of zero at $\sigma_m = 0$. This trend is

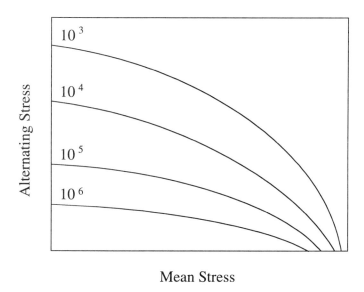

Mean Stress

**Fig. 4.4** A Haigh plot

contrary to observed test results for negative $\sigma_m$, i.e. the values of the ratio $\sigma_a / \sigma_e$ increase with increasing compressive $\sigma_m$. Equation (4.8) does exhibit the trend of increasing $\sigma_a$ with increasing compression. By including three terms of equation (4.7), as noted above, however, the effect of a compressive mean stress could be included in the curve fitting procedure.

### 4.2.4 Multiaxial stress states

The topics considered in the previous sections of this chapter focused on utilizations of the stress-life approach to problems for which uniaxial stress states are developed. In a number of service applications, however, combined or multiaxial stress states are developed. Pressure vessels under cyclic internal pressure and crank shafts under combined torsion and bending are among the examples in which multiaxial stress states occur. Although experiments in which the multiaxial stress states of concern can be duplicated, a common practice is to use available uniaxial test data which can be analytically adjusted to account for differences in the stress states. For high cycle fatigue cases the usual procedure is to apply an equivalent stress approach which is analogous to that used for predicting the transition from elastic to plastic behavior under combined stresses.

Combined stress states of two types are encountered in service. In one the ratios of the principal stresses are constant during cyclic loading. This is described as proportional loading and when two sources of loading are present they are said to be in-phase with one another. The second loading condition involves nonproportional loading, i.e. the ratio of the principal stresses varies during a cycle. Emphasis here is on the proportional loading case. The case of nonproportional loading is less well established and a consensus on the choice of a method of analysis has not been established. Elements of proposed methods of analysis for out-of-phase loading are discussed by Fuchs and Stephens (1980).

One condition for the onset of yielding can be derived in several ways which give identical results. The derivation of these equivalent methods is based on an algebraic combination of the first and second invariants of the stress tensor, the total strain energy minus the hydrostatic energy (the distortion energy), and the root mean square of the differences between the principal stresses. The resulting relationship is often described as the Mises yield condition (Johnson and Mellor 1973). In terms of principal stresses it has the form

$$\left[\left(\sigma_{11}-\sigma_{22}\right)^2 +\left(\sigma_{22}-\sigma_{33}\right)^2 +\left(\sigma_{33}-\sigma_{11}\right)^2\right]^{1/2} = \sqrt{2}\,Y, \tag{4.10}$$

where $Y$ is the tensile yield stress. Note that if only $\sigma_{11}$ is non-zero, the equation reduces to

$$\sigma_{11} = Y, \tag{4.11}$$

which represents yielding in uniaxial tension.

Often the results of stress analyses are expressed relative to a coordinate system for which the stress components are not principal values. It is not necessary to transform to principal stress values, because the Mises condition can also be written in terms of non-principal values, i.e.

$$\left[(\sigma_{11}-\sigma_{22})^2 + (\sigma_{22}-\sigma_{33})^2 + (\sigma_{33}-\sigma_{11})^2 \right.$$
$$\left. + 6\left(\sigma_{12}^2 + \sigma_{23}^2 + \sigma_{31}^2\right)\right]^{\frac{1}{2}} = \sqrt{2}Y. \tag{4.12}$$

To use equation (4.10) for fatigue analysis, it is merely necessary to replace the tensile yield stress, $Y$, by the uniaxial fatigue limit, $\sigma_e$. Also, the principal stress components are identified as the alternating stresses during reversed loading. An equation for reversed, proportional loading may then be expressed as

$$\left[(\sigma_{11}-\sigma_{22})^2 + (\sigma_{22}-\sigma_{33})^2 + (\sigma_{33}-\sigma_{11})^2 \right.$$
$$\left. + 6\left(\sigma_{12}^2 + \sigma_{23}^2 + \sigma_{31}^2\right)\right]^{\frac{1}{2}} = \sqrt{2}\sigma_e. \tag{4.13}$$

The use of this equation postulates that when the combination of stress components in the function on the left equals the quantity on the right side, the intensities of loading which lead to fatigue failure are the same for the two cases, i.e. uniaxial loading and multiaxial loading.

It should be recognized that since crack initiation usually occurs in a surface layer, the surface plane is a principal plane. Then, if principal stresses are used, the state of stress will be biaxial. If the 3 direction is taken as normal to the surface, $\sigma_{33} = 0$ and principal values are taken for the surface stresses, the condition stated in equation (4.13) simplifies to

$$\sigma_{11}^2 - \sigma_{11}\sigma_{22} + \sigma_{22}^2 = \sigma_e^2 \tag{4.14}$$

As an example of the use of these equations, consider the source of $\sigma_e$ to be a completely reversed uniaxial test. Let the multiaxial state of interest be a completely reversed torsion test. For the torsion test of the circular cylinder with the stress state shown in Fig. 4.5, the only nonzero component is $\sigma_{12}$. From equation (4.13)

$$\sqrt{6}\,\sigma_{12} = \sqrt{2}\,\sigma_e, \ or \ \sigma_{12} = \sigma_e/\sqrt{3}$$

$\sigma_{12}$ can be calculated from the equation for torsional loading, i.e.

$$\sigma_{12} = \sigma_{21} = Tr/J,$$

where $J$ is the polar, second area moment of a circular cross-section. From a knowledge of $\sigma_e$ for the given material the torque, $T$, which should give an equivalent fatigue behavior, can be determined.

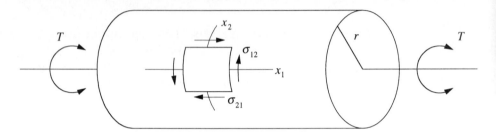

**Fig. 4.5** Torsional loading of a circular cylindrical bar

The loading conditions encountered in service are often not restricted to cases in which the principal stress values alternate about zero, i.e. they have nonzero mean values. Generalizations of equations (4.10) and (4.12) are not immediately clear.

A direct extension of equation (4.10) could be developed by simply replacing $\sigma_a$ and $\sigma_m$ by effective values based on the condition expressed in equation (4.10). Thus, it would follow that an equation of the form

$$eff\sigma_a + \frac{\sigma_e}{\sigma_u}\,eff\sigma_m = \sigma_e \qquad (4.15)$$

may be used. The term eff $\sigma_a$ would then be defined by the function on the left side of equation (4.10). Although an eff$\sigma_m$ term could also be computed by use of a Mises type function, this procedure has not been adopted for use. Experimental fatigue data have indicated that a tensile mean stress decreases fatigue life and a compressive mean stress increases fatigue life. An eff$\sigma_m$ based on a Mises type relation does not produce this effect. Sines (1955), therefore, proposed that the mean stress effect should be introduced by use of a relation based on the hydrostatic mean stress. Since the sum of the normal stresses is an invariant of the stress tensor, any sum of the normal mean stresses can be used. The mean stress term is expressed as

$$S = S_{11} + S_{22} + S_{33}$$

where the $S_{ii} = (\max \sigma_{ii} + \min \sigma_{ii})/2$

Sines' proposed equation can then be written as

$$\left[ \left( \sigma_{11} - \sigma_{22} \right)^2 + \left( \sigma_{22} - \sigma_{33} \right)^2 + \left( \sigma_{33} - \sigma_{11} \right)^2 \right]^{\frac{1}{2}} + mS = \sqrt{2}\ \sigma_e \tag{4.16}$$

where the $\sigma_{ii}$ are the alternating principal stress components, $\sigma_e$ is the uniaxial, fully reversed fatigue life for a smooth bar of the same material, and $m$ is a material constant. For a biaxial state for which $\sigma_{33} = 0$, equation (4.16) becomes

$$\left[ \sigma_{11}^2 - \sigma_{11}\sigma_{22} + \sigma_{22}^2 \right]^{\frac{1}{2}} + \frac{\sqrt{2}}{2} m \left( S_{11} + S_{22} \right) = \sigma_e \tag{4.17}$$

Note that whereas an eff $\sigma_m$ would always be positive, it follows that the term $(S_{11} + S_{22})$ may be either positive or negative. It can, therefore, reflect the effects of tensile and compressive mean stresses on the fatigue life. It should be noted that the first term on the left of equation (4.16) can be shown to be derivable from an algebraic combination of the first and second invariants of the stress tensor. The second term has the form of the first invariant. The left side of equation (4.16) can, therefore, be recognized as an algebraic combination of the first and second invariants of the stress tensor (Johnson and Mellor 1973).

The application of multiaxial stress criteria for fatigue under proportional loading and for proposed extensions to nonproportional loading are described by Fuchs and Stephens (1980).

Some forms of fatigue in which multiaxial stress states are encountered are not readily analyzed by the preceding methods. These include rolling fatigue and fretting fatigue. Rolling fatigue can occur on surfaces in contact with ball and roller bearings, cams and gearing. These involve contact between curved surfaces which are repeatedly brought together and produce high localized stress states. The stress states which lead to rolling contact fatigue failure are usually triaxial with a large hydrostatic compressive component. The maximum shear stress below the contact zone is about three times higher than that at the surface and cracks below the surface are nucleated. These cracks can grow and develop into chips which can break away from the surface.

Rolling fatigue, therefore, represents an example in which crack initiation does not begin at the surface. The mechanism in which chips are formed is often described as spalling or flaking. Empirical relations in which contact force and life cycles are used to provide a basis for design are available (McClintock and Argon 1966). The type of lubricant used has a significant effect on lifetime. Morrison (1955) has shown that lifetimes can differ by a factor of up to 60 for different lubricants. As would be expected, the choice of a lubricant must also consider the possibility of coupled corrosion-fatigue mechanisms.

Fretting fatigue originates between two contacting surfaces which slide relative to one another. The sliding displacements are oscillatory in character and the amplitudes of the displacements are restricted and very small. Fretting can occur in pin type connections, in pressed fittings and, in fact, between any two surfaces subject to oscillatory loading for which some relative sliding displacement is present. Coupled fretting corrosion fatigue must also be recognized as an issue in design. Data on strength reduction factors for fretting fatigue contact between dissimilar metals are given by Waterhouse (1972). A review of issues which have been identified in fretting fatigue has been presented by Lindley (1992).

### 4.2.5 Environment

Environmental effects which can affect fatigue behavior are extremes in temperature and active or corrosive atmospheres.

Exposure to very low temperatures can result in an increase in the endurance limit. Since many metals exhibit a significant reduction in fracture toughness at low temperatures, however, the introduction of fatigue cracks can cause these metals to be vulnerable to brittle fracture. The brittle response of body centered cubic metals such as the ferritic alloys below their transition temperature is the prime example of this behavior. The widespread use of ferritic alloys in pressure vessels, bridges and ships has attracted considerable attention to reported cases of brittle fracture in these nominally ductile alloys. Studies prompted by these problems have resulted in recommendations focusing on design details involving reductions of stress concentrations,  developments of strategies for crack arrest, and improved material selection (Parker, 1957).

The use of metals in engines introduces the other extreme, i.e. elevated temperature exposure. As operating temperatures approach and exceed about one half of the homologous temperature ratio, time dependent deformation, or creep, becomes a factor which must be considered. The homologous temperature ratio is defined as the ratio of the absolute exposure temperature to the absolute melting temperature of a given metal. For lead this occurs very close to ambient or room temperature. For the heat resisting alloys used in gas turbine engines the operating temperatures are, of course, much higher.

Cyclic loading at elevated temperatures can be viewed from both phenomenological and mechanistic perspectives. From a phenomenological perspective the frequency of cycling and the shape of the loading cycle can result in quite different responses. This is illustrated in Fig. 4.6. The loading frequencies in Fig. 4.6 (a) and (b) are the same, but the time spent at the maximum load, the dwell time, is much greater in (a) than in (b). Since creep is a time-dependent phenomenon, the behavior of the two loading histories can be expected to be quite different.

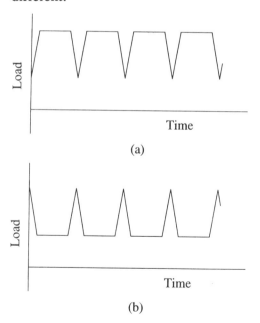

**Fig. 4.6** Forms of loading cycles

A mechanistic examination indicates that creep and fatigue mechanisms are quite distinct. As noted previously, fatigue is the result of crack nucleation in localized slipbands within individual grains which are located in a surface layer. Creep deformation mechanisms develop within the interior of a body and are dependent on the increased elevated temperature mobility and diffusion of vacancies, atoms and dislocations (Hertzberg, 1989). If both of the mechanisms associated with fatigue and creep are occurring simultaneously, it is reasonable to expect that the mechanisms can interact with one another. That is, they can be expected to involve coupled phenomena which are not simply additive. Although the impact of creep mechanisms on fatigue crack nucleation is of interest, there is more engineering interest in coupled fatigue – creep crack growth behavior. This is discussed in Chapter 5.

A good introduction to corrosion in metals and plastics is presented in a text by Heitz, Henkhaus and Rahmel (1992). Also, a variety of corrosion fatigue problems are considered in conference proceedings edited by Scott and Cottis (1990).

The effect of active atmospheres on stress-life behavior has been the subject of research studies for a number of years (Forrest, 1962). In corrosive atmospheres a fatigue limit or endurance limit does not exist. Even an atmosphere of air can influence fatigue life. Tests in a vacuum have, in fact, exhibited substantial increases in lifetime over those observed for tests in air. The coupled corrosion – fatigue behavior resulting from cyclic loading and corrosive attack has been recognized as a particularly important problem in aircraft. As with of creep – fatigue, however, the dominant concern is with the tolerance of metals to fatigue crack growth in the presence of active or corrosive atmospheres. This topic is discussed in section 5.7.4.

### 4.2.6  Variable amplitude loading

The components in stationary machines and engines are often subjected to cyclic loading for which the amplitudes of the mean and alternating loads are nominally repetitive during their operation. Data from reversed, constant amplitude loading tests and the schemes for extending the use of these data for nonzero mean load and biaxial state cases can be utilized for engineering design projects. In nonstationary vehicles, however, the amplitude of loading varies with time and different analytical procedures are required. This type of load exposure is aptly described as variable amplitude loading. A prime example of this type of loading occurs in aircraft and the loading spectra are not only complex, but differ for different types of aircraft, e.g. fighter aircraft, transport aircraft. Thus the differences in spectra depend upon usage and mission (section 2.2 and ten Have, 1989).

Proposals by Palmgren (1924) and Miner (1945) attempted to account for variable amplitude loading effects by the use of the assumption that the fatigue damage incurred during each cycle was independent of the prior loading history. The rule based on this assumption is described as the linear damage rule. If the amplitude of loading is $\sigma_1$ for $n_1$ cycles and the fatigue life at $\sigma_1$ is $N_1$, the fraction of life used would then be $n_1/N_1$. When this ratio equals unity, fatigue failure is said to occur.

For two loading amplitudes, $\sigma_1$ and $\sigma_2$, the damage accumulated is

$$\frac{n_1}{N_1} + \frac{n_2}{N_2}. \tag{4.18}$$

The values of $n_1$ and $n_2$ which cause the sum of equation (4.18) to be unity represent the combinations of loading which would be predicted to lead to fatigue failure.

In actual applications the ratio of $n_1$ to $n_2$ can be determined by analysis or experiment, so the cycles to failure could be predicted. The basic idea embodied in equation (4.18) can be extended. It states that fatigue failure occurs when

$$\sum_{i=1}^{k} \frac{n_i}{N_i} = 1, \tag{4.19}$$

where $k$ is the total number of cycles in a loading spectrum,
$\quad\quad i$ is the $i$th applied stress level,
$\quad\quad n_i$ is the number of cycles at stress level $\sigma_i$,
$\quad\quad N_i$ is the fatigue life at stress level $\sigma_i$.

Although the linear damage rule is intended to provide a basis for predicting fatigue life under variable amplitude loading, it has deficiencies which should be recognized. Experiments have indicated that the sum of equation (4.19) can, depending upon the order in which load levels are applied, be either greater or less than unity. If, for example, a block of high level loading is followed by a block of low level loading, the sum for failure in equation (4.19) can be less than unity. For notched specimens the reverse, however, is found to occur. If the block order is changed to a low – high sequence, the sum for failure is greater than unity. It has been argued that if the loading spectrum is a mixture of high – low and low – high sequences, the deviations from unity may be canceled. This does not, however, constitute a generally sound basis for predictions. Buch, Seeger and Vormald (1986) have proposed a simple procedure for correcting for the deviation of the right hand side of equation (4.19) from unity. They suggest that if a newly considered load spectrum is similar to that of a spectrum for which the deviation is known, a corrected estimate of life, $N''$, can be obtained by multiplying the newly calculated life, $N''_{calc}$, by the ratio for the previous, similar spectrum, $N'_{act}/N'_{calc}$. This ratio can of course be either greater than or less than unity. Experience would naturally be a prerequisite for judging the similarity of the two loading cases. Other alternative methods for assessing the accumulation of damage under variable amplitude loading have been proposed, and some of these have been discussed by Collins (1981). The introduction of material constants and the complexity of applying them to complex spectra, however, often detract from their usefulness.

As noted previously, there is an increasing emphasis in design on determining the tolerance of components to existing cracks. The subject of variable amplitude loading is, therefore, again discussed in Chapter 5, in which fatigue crack growth is discussed.

## 4.3 THE STRAIN-LIFE STRATEGY

### 4.3.1 Introduction

The smallest number of cycles in a stress-cycles to failure test can be determined by conducting a fully reversed uniaxial test on a smooth bar. If the cycle begins with the tensile excursion, and the load is increased until the ultimate strength is reached, the cyclic failure number would be 0.25, i.e. only one quarter of a full cycle would be achieved. Smaller maximum loads could, of course, provide data for complete cycles to failure. The range of life cycles designated as low cycle fatigue varies, but can extend to about $10^5$ cycles. It is within this range that design procedures based on a strain-life strategy are used.

Although uniaxial smooth bars are typically used to generate data for strain-life analyses, the test specimens do not represent the geometrical configurations encountered in service components. The critical sites in these service components are located at notches, holes and fillets, and they involve enclaves of plastically deformed material which are constrained within an elastically strained body. The elastic constraints introduced at these locations result in cyclic strains which are more readily determined than the corresponding stresses. Because of the preferred reference to cycle strains, low cycle fatigue data are generated in strain controlled tests rather than the stress controlled tests which were described for high cycle fatigue.

For a single, complete strain cycle of the type depicted in Fig. 2.4 the stress – strain response for loading into the plastic range is illustrated in Fig. 4.7. Note that the initial response after loading from the origin, and the initial response after load reversals at P and Q are linear and elastic. The stress value at point Q is often, however, observed to be less in magnitude than that at point P due to the Bauschinger effect (1886). The total strain range $\Delta \varepsilon_t$ is given by the equation

$$\Delta \varepsilon_t = \Delta \sigma / E + \Delta \varepsilon_p \ , \tag{4.20}$$

where $\Delta \sigma$ is the stress range,
  $\Delta \varepsilon_p$ is the plastic strain range, and
  $E$  is Young's elastic modulus.

The area within the loading cycle represents the work done on the material. Some of this work is absorbed in internal structural changes and some is converted to heat. In metals this results in cold working, and in polymers in molecular chain realignments.

The type of continuing response exhibited by different materials can vary significantly. For a constant $\Delta \varepsilon_t$ the $\Delta \sigma$ for some metals can continuously increase with increasing strain cycles. This phenomenon is described as cyclic hardening. For others the $\Delta \sigma$ can continuously decrease. This is described as cyclic softening. The hardening and the softening tend, however, with continued cycling to achieve a stable state, i.e. the enclosed loading loops, which have been changing continuously, develop a form which is repeated with further cycling. A cyclic stress – strain curve which differs from the monotonic stress – strain curve can be obtained by plotting a curve through the tips (such as P and Q) of the stress – strain plots for continuing cycles with increasing values of $\Delta\varepsilon_t$.  A stable stress – strain curve must be achieved for each $\Delta\varepsilon_t$ value.

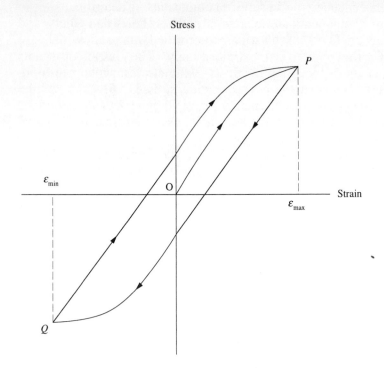

**Fig. 4.7**   Stress – strain plot for a single strain cycle

Based on comparisons of the ratios of ultimate strength to 0.2 per cent offset yield strength, Manson and Hirschberg (1964) have suggested that initially soft metals tend cyclically to strain harden and initially hard metals tend to soften.  Metals with ultimate strength to yield strength ratios of greater than 1.4 harden.  Those for which the ratio is less than 1.2 exhibit softening.  For values of this ratio between 1.2 and 1.4 changes in hardness can be expected to be small.  Hertzberg (1989) has presented a tabulation in which the properties developed by cyclic straining are summarized for selected ferrous, aluminum, copper and nickel alloys.

Cyclic strain experiments performed on amorphous and crystalline polymers (Hertzberg, 1989) indicate that only cyclic softening occurs. This is consistent with the reaction to hysteretic heating, i.e. work transformed to heat reduces the resistance to continuing deformation (see section 3.3).

The preceding discussion has been confined to a description of behavior in which cyclic straining has been imposed on an initial condition for which the mean stress is zero. If the initial mean stress is tensile, a continuing extension can occur. If the mean stress is compressive, a contraction can occur. This behavior has been described as a cyclic strain induced creep.

### 4.3.2 The strain-life diagram

An empirical relationship between cyclic strain and the number of cycles to failure for cyclically stabilized metals can be developed by partitioning the total strain into elastic and plastic components. The elastic component, $\Delta\varepsilon_E$ has been found to obey the following relationship which is attributed to Basquin (1910).

$$\Delta\varepsilon_E = 2\frac{\sigma_a}{E} = 2\frac{\sigma_f'}{E}\left(2N_f\right)^b, \qquad (4.21)$$

where $\Delta\varepsilon_E$ is the elastic strain range,
$\sigma_a$ is the true stress amplitude,
$\sigma_f'$ is a fatigue strength coefficient,
$N_f$ is the number of cycles to failure,
$E$ is Young's modulus of elasticity, and
$b$ is a fatigue strength exponent.

$\sigma_f'$ corresponds approximately to the true fracture strength and $b$, a material property, has been found to vary between –0.05 and –0.12. The quantity $2N_f$ is the number of strain reversals.

Manson (1954) and Coffin (1954) independently found that the plastic strain range, $\Delta\varepsilon_p$, can be represented by the relation

$$\Delta\varepsilon_p = 2\varepsilon_f'\left(2N_f\right)^c, \qquad (4.22)$$

where $\varepsilon_f'$ is a fatigue ductility coefficient, and $c$ is a fatigue ductility exponent.

$c$ is a material fatigue property which has been found to vary between – 0.5 and – 0.7 and $\varepsilon_f'$ corresponds to the true fracture ductility, i.e. the natural or true strain at fracture.

These two empirical equations can be added to give an equation for the total strain range, $\Delta\varepsilon_T$. Thus,

$$\Delta\varepsilon_T = 2\frac{\sigma_f'}{E}\left(2N_f\right)^b + 2\varepsilon_f'\left(2N_f\right)^c. \tag{4.23}$$

The log – log plot of Fig. 4.8 provides a qualitative description of the three strain range quantities of equations (4.21), (4.22) and (4.23). The shapes of the hysteresis loops change with increasing numbers of cycles. The widths and the enclosed areas of the loops decrease as $N_f$ increases.

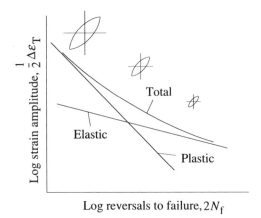

**Fig. 4.8**  Construction of a strain-life diagram

An interpretation of failure in the strain-life representation can vary from the complete fracture of a specimen to the development of a crack of an agreed upon length. In choosing a failure criterion it is as well to note that the data on smooth bars must be reconciled with failure in a component at, for example, a hole or a notch.

The form of the strain-life curves differs for different metals and extremes are observed for hard and soft metals. This is qualitatively illustrated in Fig. 4.9 for three hypothetical metals. The intersection of the curves has been observed to occur at a strain amplitude value of about 0.02 and a cycles reversal value of about 2 x 10³. For large values

of $\Delta\varepsilon_T$ the soft metal has the longest life. For smaller values of $\Delta\varepsilon_T$ the hard metal is best.

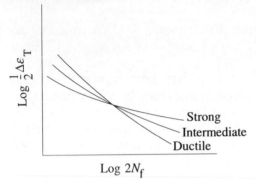

**Fig. 4.9**  Variations in strain-life diagrams

Reference to the effect of notches is made in section 4.2.2 as a part of the discussion of the stress-life strategy. The stress concentrations described in that section are valid only if the applied loads do not produce localized yielding adjacent to the root of the notch. When yielding occurs, the ratio of the local stress to the nominal stress is less than the value of the elastic stress concentration factor and it decreases with increasing applied load. In the elastic range the ratio of the local strain to the nominal strain is equal to the elastic stress concentration factor. As yielding proceeds, however, this ratio increases. Thus, whereas the strain ratio increases, the ratio for stresses decreases.

Neuber (1961) has developed an analysis of notches which provides an effective stress concentration factor which can be used to determine the cyclic strain at the notch root. The cyclic strain at the notch root can then be used in equation (4.23) to determine the number of cycles to failure. Neuber's effective stress concentration factor, $K_e$, is given by the relation

$$K_e = \sqrt{K_\sigma K_\varepsilon} \tag{4.24}$$

where $K_\sigma$ is the actual, decreasing stress concentration factor and $K_\varepsilon$ is the actual strain concentration factor. Details of the computational procedure used for notch strain-life analysis are provided by Bannantine, Comer and Handrock (1990) and by Fuchs and Stephens (1980).

The form of Neuber's rule based on equation (4.24) can be used when the amount of yielding is small. Seeger and Heuler (1980) have proposed a modified version of Neuber's rule for cases in which extensive yielding has occurred.

For variable amplitude loading the Palmgren – Miner rule can be used. The strain-life curve is then used instead of the stress-life curve described in section 4.2.6.

The effect of nonzero mean loads has been investigated and experimental results have indicated that tensile mean loads are detrimental and that compressive mean loads are beneficial (Bannantine, Comer and Handrock 1990). It should be noted, however, that compressive overloads can under some conditions prove to be detrimental. This behavior is discussed in section 5.3.5.

The types of loading histories that can be encountered in service components can be considerably more complex than the cyclic strain histories which are used to generate the strain-life diagrams of the type depicted in Figs 4.8 and 4.9. Bolt holes in gas turbine disks, for example, are sites at which low cycle fatigue conditions are developed. In addition to centrifugal forces which generate stresses that vary with engine rotational speed, time varying temperature gradients introduce thermal stresses. Data for cyclic straining at service temperatures must, therefore, be analyzed on a cycle by cycle basis to trace the accumulation of cyclic damage. The codes developed for this type of analysis must incorporate details of flight missions and they are much more complex than those used for constant amplitude strain cycling.

## 4.4 STATISTICAL CONSIDERATIONS

### 4.4.1 Introduction

The inherent scatter exhibited in the fatigue data which are needed for the application of the stress-life design strategy requires the use of statistical methods. A variety of factors contribute to the observed scatter. These include specimen location in the material stock, slight differences in heat treatment or specimen surface condition, and sampling which may include materials processed at different times.

Frost, Marsh and Pook (1974) have also discussed another basic factor which affects the distribution curves for each stress level. This involves the roles of the crack initiation and the crack growth phases of fatigue. They observed that the distribution curve obtained in tests is

composed of a distribution curve for crack initiation and a distribution curve for crack growth. It is generally accepted that the scatter for crack initiation is greater than that for crack growth. The distribution curve for the total life may, therefore, have a bimodal form.

At high stress levels the number of cycles required for crack initiation is small, so the crack growth distribution dominates and scatter is small. At low stresses crack initiation dominates and results in wide scatter. At intermediate stress levels both mechanisms make significant contributions, and a two-humped, bimodal distribution is sometimes observed.

Frost, Marsh and Pook (1974) have also observed that in fatigue tests on multi-component structures, failure may occur in different components at different stress levels. Thus, in riveted sheet construction they observed that rivet failures occur at high stress levels and sheet failures occur at low stress levels. The stress-life curve evolves, therefore, from two distinct distributions at the stress level extremes. In the transition region both modes of failure can occur and the resulting distribution is observed to be skewed, i.e. it is not symmetric about the mean value. Since multi-component construction is the norm in service applications, the ramifications of multiple site fatigue damage must be recognized as an important problem. Multiple site damage is discussed in section 6.3.

## 4.4.2 Statistical distributions

Often a Gaussian or normal distribution is used to provide an analytical representation of fatigue data at a given stress level. It has the form

$$p(x) = \frac{1}{\sqrt{2\pi}\,\sigma} e^{-\frac{1}{2}\left(\frac{x - x_o}{\sigma}\right)^2} \tag{4.25}$$

where $x_o$ is the mean value and $\sigma$ is the standard deviation. As applied in fatigue analysis,

$$x = \log N \tag{4.26}$$

where $N$ is the number of cycles. This is called the log normal distribution.

Weibull (1961) has proposed a distribution function which is frequently used in the analysis of fatigue data.

It has the form

$$p(N) = \frac{b}{N_a - N_o} \left( \frac{N - N_o}{N_a - N_o} \right)^{b-1} e^{-\left( \frac{N-N_o}{N_a-N_o} \right)^{b}} \tag{4.27}$$

where $N_a$, $N_o$ and $b$ are constants to be computed from experimental data.

A two parameter form of the Weibull distribution function can be extracted from equation (4.27) by choosing the minimum number of cycles to failure to be zero. This corresponds to setting $N_o = 0$.

The probability of failure for a given value of $N$ can be determined by integration of the distribution function for each stress level for which data are available. A qualitative representation of stress-life probability curves is shown in Fig. 4.10. The labels indicate the percent of failures that would be expected for a given curve.

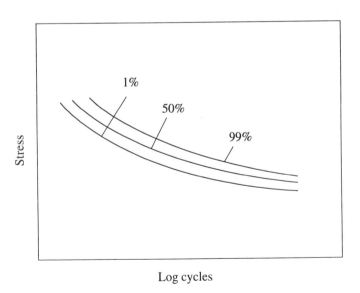

**Fig. 4.10** Scatter in terms of probability of failure

### 4.4.3 Reliability analysis

If the value of the stress to be encountered in a structural component $\sigma_L$ is precisely known, and the value of the failure stress, $\sigma_F$, of the component material is exactly known, it is clear that an acceptable design must require that $\sigma_L < \sigma_F$. If this inequality is not satisfied, the design of the component must be modified to reduce $\sigma_L$. The conditions cited in this idealized example are, however, atypical. Often, the loads on components vary with time as with variable amplitude loading. Also, the mechanical properties vary from sample to sample.

If an analysis of a loading spectrum is performed, a statistical distribution of the stress in components can be determined. Similarly, a distribution of the relevant mechanical properties can be established. When the two distributions are well separated, as shown in Fig. 4.11 (a), the situation resembles the idealized example described above. When they overlap, as in Fig. 4.11 (b), there clearly is a possibility of failure in some components.

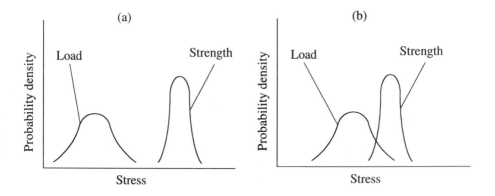

**Fig. 4.11** Probability density functions for stress due to loading and for failure stress

The objective of reliability analysis is to provide a formal basis for determining the probability of failure due to the overlapping of the load and strength distributions. The loading spectra encountered in aircraft and the uncertainties associated with fatigue phenomena provide an example in which reliability analysis is both appropriate and expedient.

Reliability, as applied to engineering, is defined as the probability that there is no failure, i.e. one minus the probability of failure (O'Conner, 1988). As applied to a major component in a fleet of aircraft, it requires the availability of a load spectrum which is scaled to give the loads and hence stresses in the component of interest. These loads, in turn, must be adjusted to provide equivalent, zero mean stresses by use of, for example, a Goodman diagram (section 4.2.3). The results form the basis for determining a density or distribution function for loads. Distribution functions must then be determined for strength from stress versus cycles to failure data, i.e. the variation in strength at each value of cycles.

A determination of reliability can then be made either by an analysis of the joint or bivariate distributions or by evaluation of a safety margin parameter (O'Conner, 1988) which is expressed in terms of mean values and standard deviation values of the strength and load distributions. Also, for the variable amplitude loading implied by the load distribution function a damage rule such as the Palmgren – Miner rule for cumulative damage may be introduced (section 4.2.6).

An application of reliability analysis with an emphasis on helicopters has been described by Everett, Bartlett and Elber (1990). A goal of these analyses is to establish design and operational procedures which can be used to estimate the risk of failure. For example, how many failures could be expected in a component of an aircraft which is part of a fleet after a given number of missions? An operative application of this methodology can have an impact on choices of inspection intervals and on the replacement of components.

The application of damage tolerance concepts to risk management of fleets of aircraft is discussed in section 6.2.1.

## 4.5 NONMETALLIC MATERIALS AND COMPOSITES

Efforts designed to take advantage of new technological developments often result in attempts to develop extensions or modifications of engineering solutions which have proven to be useful in the past. As interest in the use of improved nonmetallic materials and composites has

increased, examination of the adaptability of failure criteria originally developed for metals has been the objective of numerous studies. In the discussion which follows, plastics, brittle solids and composites with metal, polymeric and ceramic matrices are considered, and the extent to which the safe life strategies used for metals are applicable is examined.

### 4.5.1  Polymeric solids

A cursory examination of fatigue data on plastic, smooth bar specimens would appear to indicate that design strategies based on either stress or strain versus cycles to failure data would be acceptable. A review of the fatigue behavior mechanisms described in section 3.3, however, reveals that any design procedure which is adopted must include a consideration of possible dependencies on frequency, size and geometry. Also, the effects of mean stress and biaxial stress states cannot be neglected.

For metals there are methods available for extending the use of standard test data to account for the conditions to be encountered in service. Goodman type diagrams which account for mean stress effects are examples of this type of extension. For plastics, however, acceptable analytical adjustments of data from so-called standard tests for which the test conditions differ from those for a projected service application are difficult to achieve.

Standardized tests are of value in identifying failure mechanisms for a variety of test conditions. They also can be useful as a screening procedure for material selection. For design purposes, however, specialized testing is indicated. If service conditions are duplicated, data on load versus cycles to failure can be an important ingredient in the evolution of a design. This suggests that for plastics the practice of component fatigue tests initiated in 1860 still can provide a viable basis for sound design.

### 4.5.2  Brittle solids

The quality of brittle ceramics has improved considerably in recent years. The research which has been responsible for the improvements and the identification of failure mechanisms are reviewed in section 3.4. Despite the progress which has been made, it may be concluded that two features of brittle materials limit their applicability in structural components. These are an inherent intolerance to flaws and cracks and a size and stress state dependence on fracture. Some applications involving loading which is dominantly compressive and/or very high temperature environments are, however, feasible. The characteristics of the data which are available on fatigue due to cyclic loading do not, however, provide support for the application of the type of safe life design strategies used for metals. The uncertainties which stem from fabrication quality (initial flaw distribution) and the statistical effects of size and stress state lead to the conclusion that brittle materials should be evaluated by duplicating the expected service conditions. Thus, although controlled standard tests can be useful in evaluating and screening candidate materials, component tests should be used as a basis for evolving

### 4.5.3 Composites

Fatigue crack initiation mechanisms in composites are dependent on matrix-reinforcement details and loading. This topic is discussed in sections 3.5, 3.6 and 3.7. Because of the extensive variety of matrix-reinforcement combinations which have been developed, it is difficult to offer generalizations on the adaptability of this class of materials to the use of safe life design strategies. When they can be used, moreover, test programs must be conducted on each specific composite. It may be anticipated that testing may include the effects of mean stress, tensile versus compressive loading, loading orientation relative to fiber axes and stress concentrations. The details of the attachment of composite components to adjacent components must also be considered (Agarwal and Broutman, 1990).

As observed in section 3.7, the use of ceramic matrix composites is of primary interest in elevated temperature applications. Both monotonic and cyclic loading can lead to progressive degradation and the sequence of events has a statistical character. Fatigue data which have been obtained indicate that these materials are not damage tolerant, and design methods especially tailored to account for coupled fatigue damage and creep need to be adopted.

Fatigue in metal matrix composites is considered in section 3.5. In addition to the bonding together of metal sheets to form laminates, there are a number of combinations of composites which are reinforced by the use of continuous fibers, short fibers and particles. A variety of fabrication techniques and reinforcement combinations are employed (section 3.5.1).

Some of the fabrication techniques have been developed to the degree that products are commercially available. These include the laminated aluminum alloy sheets referred to in section 3.5.2 and to heat treatable aluminum alloys which are reinforced by silicon carbide fibers. These latter products are available in the form of billets, plates, sheets, extrusions and forgings and they exhibit mechanical properties which are quite attractive. The tensile elastic modulii are approximately 50% greater than those for unreinforced aluminum alloys and their tensile strength is comparable with that of commonly used aluminum alloys. Stress corrosion cracking can also be averted by an appropriate selection of the matrix alloy.

Failure mechanisms for different reinforcement elements and for loading states are discussed in section 3.5.2. Although metal matrix composites have a potential as replacements in applications in which metals have been used, it can be anticipated that extensive testing will be required to establish a basis for the use of safe life design procedures. This may include tests on the effects of stress ratio, stress concentration, and multiaxial stress states. Data on these effects must be obtained for each matrix-reinforcement combination for both low and high cycle fatigue.

If sufficient test data are available for a given metal matrix composite, design envelopes which incorporate a knowledge of failure mechanisms may be constructed for high cycle fatigue. This is illustrated in Fig. 4.12 by a plot of alternating stress, $\sigma_a$, versus mean stress, $\sigma_m$, for a unidirectionally reinforced composite. For loading parallel to a continuous fiber reinforcement, tensile and compressive load limits can be anticipated. The curve in the right quadrant represents either an endurance limit loading, or failure after a given number of cycles. Two types of compressive failure can be developed. These are kinking, which is a localized form of instability discussed in section 3.5.2, or Euler buckling. The latter depends upon the slenderness ratio. As the slenderness ratio decreases, that line would be translated to the left. Note that for the example of Euler buckling shown, $\sigma_e > |\sigma_c|$. For the plot shown this implies that buckling may occur for positive $\sigma_m$. This may occur for a 'long' column.

Kinking is a localized form of inelastic buckling. Although the boundaries for the negative $\sigma_m$ mechanisms are shown as straight lines, the actual shapes and locations on a diagram would have to be determined for each metal matrix composite. As shown they merely illustrate how the identification of failure modes may be used to construct design curves.

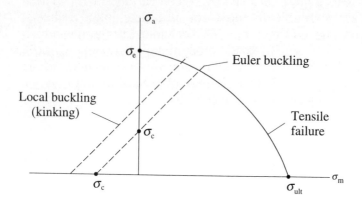

**Fig. 4.12**   Metal matrix failure diagram

For multiaxial stress states criteria such as that proposed by Sines (section 4.2.4) would have to be generalized to account for anistropy. Failure mechanisms resulting from the presence of holes or notches must be identified for each metal matrix composite. Also, versions of the strain-life methods discussed in section 4.3 need to be developed for low cycle fatigue applications. These general observations reveal that though much fatigue data have been obtained on metal matrix composites, the objectives of the investigations were to characterize this class of materials and to identify failure mechanisms. The programs were generally not created to provide standard design data. It is also of interest to observe that there is a potential for extending the useful temperature range of a given alloy by the use of reinforcing elements. The coupling of creep and fatigue in applications which may take advantage of this type of composite can be expected to involve mechanisms which cannot, however, be adequately incorporated in the application of the safe life strategies which have been discussed here.

The fabrication and properties of polymeric matrix composites are discussed in sections 3.6.1 and 3.6.2. The mechanisms which lead to damage in these composites include debonding of the interface between fibers and matrix and interlaminar microcracking. As noted in section 3.6.2, cross-ply layups are particularly vulnerable to delamination damage. The origins of failure differ with different layups and for different types of loading. Failures in unidirectionally reinforced composites are, for example, strongly dependent upon the orientation of loading. Also, mechanisms of failure differ for tensile and compressive loading.

Agarwal and Broutman (1990) have discussed the effect of mean stress on fatigue failure and have described the use of a Goodman type diagram which has been proposed by Boller (1957). Goodman diagrams display the effect of tensile mean loads by straight lines (section 4.2.3). The plot shown in Figure 4.13 is analogous to a Gerber type diagram (section 4.23).

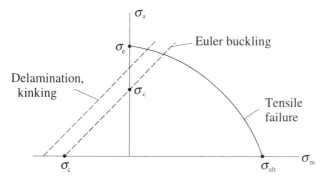

**Fig. 4.13** Polymeric matrix failure diagram

Polymeric matrix composites are also vulnerable to damage mechanisms produced by compressive loading. Two compressive damage modes have been indicated in Fig. 4.13. As for metal matrix composites, a possible failure mode is Euler buckling. A second compressive failure mode is delamination which is initiated by localized buckling. A third mode of failure involves a localized kinking which can, for brittle fibers, result in fiber fracture (section 3.6.2). This last failure mode is not depicted in Fig. 4.13.

Despite the wide variety of products which are available, considerable experience in their use has been accumulated, and with appropriate experimental evaluation polymeric composites can be used with confidence in structural applications. Mechanisms of failure have been identified and techniques for monitoring components have been developed. Also, the reduction in strengths of these composites due to fatigue can be substantially less than those for their metal counterparts.

# 5

# Fatigue crack growth

## 5.1 INTRODUCTION

For many years safe life strategies have played a dominant role in the design of machines and structures which are subject to fatigue. They can, moreover, be expected to be useful in many future design projects. There is no likelihood that they will be completely superseded by different methodologies. During the 1950s, however, while the theory of linear elastic fracture mechanics was being developed, an expanding interest in the growth of cracks under cyclic loading evolved. The growth of these cracks was recognized as a stable process, i.e. growth occurs over a number of loading cycles instead of proceeding abruptly to complete fracture.

Cracks in solids can be present as a consequence of manufacturing processes or they can be nucleated as the initial phase of the fatigue process. Cracks can, therefore, be present in machines and structures which are in service. Clearly, then, it is logical to enquire about the degree to which a machine or structure can tolerate the presence of such cracks under given service conditions.

## 5.2 IMPLEMENTATION OF A CRACK GROWTH STRATEGY

The determination of how many cycles of loading can be applied before a critical condition is reached requires the availability of an equation for determining crack growth. Also, the methodology implied indicates that sites which have been identified as critical should be monitored by the use of non-destructive testing instruments. The elements of the concepts outlined form the basis of a method which evaluates damage tolerance. Structural integrity is then judged by the extent to which tolerance to damage is achieved.

In view of the scatter associated with the collection of data for the stress-life method, concern about comparable difficulties might be anticipated in an implementation of a crack growth strategy. It has, however, been observed by Frost, Marsh and Pook (1974) that the contribution to total scatter of the crack growth phase in a fatigue test is much less than that for the crack initiation phase. Pook (1983) has summarized research which has been conducted to evaluate the potential scatter in the crack growth lifetime.

### 5.2.1  Rate of crack growth equations

The initial goal of investigators was to discover a functional relationship which described crack growth rate. Initially, it was assumed that the rate of growth is proportional to the applied stress and the crack length. In functional form this resulted in the equation

$$\frac{da}{dN} \propto F(\sigma, a), \tag{5.1}$$

where $da/dN$ is the cyclic rate of crack growth, $\sigma$ represents an externally applied traction in the form of a stress, and $a$ is the crack length.

When a power relation was selected for the function $F$, equation (5.1) was written as

$$\frac{da}{dN} \propto \sigma^m a^n. \tag{5.2}$$

Frost (1959) analyzed laboratory results and found that values of $m$ and $n$ of about 3 and 1, respectively, could be used in equation (5.2) to represent test data. Liu (1963) proposed that the range of stress, $\Delta\sigma$, should be substituted for $\sigma$ in equation (5.2), and he used an energy based argument to suggest values of $m = 2$ and $n = 1$.

Although forms of the crack growth rate relation of equation (5.2) can be used to correlate data obtained from laboratory experiments, they are not suitable for application to the variety of loading conditions and component geometries encountered in service. What would, of course, be desirable for such applications is a parameter which incorporates the loading details and the geometry of the cracked body. The mode I stress intensity factor provides this type of representation for linear elastic fracture mechanics and this led Paris, Gomez and Anderson (1961) to explore its use as the parameter for describing fatigue crack growth. They suggested a crack growth rate law which was a function of the maximum stress intensity factor during cycling, and the ratio of the minimum stress to the maximum stress. Using fatigue crack growth data on 2024-T3 aluminum sheet from three different laboratories, they demonstrated that the use of the stress intensity factor provided support for their conjecture. Donaldson and Anderson (1962) subsequently provided additional support for the use of the mode I stress intensity factor by demonstrating successful correlations for two aluminum alloys and two steels.

The driving force in stress-life correlations had early been found to be the range of applied stress. In a subsequent paper Paris and Erdogan (1963) suggested a crack growth equation of the form

$$\frac{da}{dN} = C(\Delta K)^n, \tag{5.3}$$

where $\Delta K$ is the range of the mode I stress intensity factor. For the data they displayed they found that $n = 4$ fitted the results in the mid-range of crack growth rate. This equation has subsequently been described as the Paris law.

The qualitative features of crack growth rate versus stress intensity factor range on a log – log plot are illustrated in Fig. 5.1. The data required to generate the mid-range lines shown are obtained from tests in which the cyclic stress range, $(\sigma_{max} - \sigma_{min})$ is constant. Note that as the crack length changes, $\Delta K$ changes.

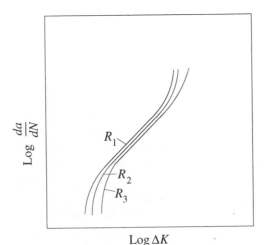

**Fig. 5.1** Crack growth rate versus mode I range of stress intensity factor for ratio values of $R_1 > R_2 > R_3$

Distinct curves are observed for different values of the stress ratio, $R$. As shown in Fig. 5.1, the growth rate tends to increase with increasing $R$ for a given $\Delta K$. It should, however, be noted that a reversal of this trend has been observed for negative $R$ values. This behavior is discussed in section 5.3.5.

The lower parts of the curves exhibit a threshold value of $\Delta K$ below which crack growth is essentially arrested. The rate of growth in this vicinity is of the order of $10^{-8}$ mm/cycle. A discussion of the threshold behavior is presented in section 5.2.2 The upper parts of the curves represent the response developed as the $K_{max}$ value approaches the critical stress intensity factor $K_{IC}$. The rate increases without bound as $K_{max} \rightarrow K_{IC}$.

Clearly, the complete curves of Fig. 5.1 are not represented by the Paris equation. The upper part of the curve depends on $K_{IC}$ and the lower part depends on a threshold value, $\Delta K_{th}$. An equation which incorporates all of the features shown in Fig. 5.1 has been used in the NASA/FLAGRO computer program for mode I fatigue crack growth (NASA/FLAGRO, 1989). It has the form

$$\frac{da}{dN} = \frac{C(1-R)^{m}\Delta K^{n}\left(\Delta K - \Delta K_{th}\right)^{p}}{\left[(1-R)K_{IC} - \Delta K\right]^{q}}. \tag{5.4}$$

This equation can exhibit all of the features shown in Fig. 5.1, i.e. the threshold effect, the $R$ ratio effect and the final instability event. Values of fatigue thresholds have been compiled in a book by Taylor (1985).

Equation (5.4) reduces to other equations which have been proposed as follows:

| Exponent values | Equation |
|---|---|
| $m = p = q = 0$ | Paris |
| $m = p = 0, q = 1$ | Forman |
| $p = q = 0, m = (m_w - 1)n$ | Walker |

The onset of instability predicted by equation (5.4) can be more clearly realized by rearranging the denominator. It can be shown that the denominator can be rewritten as

$$\left[\frac{\Delta K}{K_{max}}\left(K_{IC}-K_{max}\right)\right]^q.$$

(5.5)

In this form it can be seen that as $K_{max}$ approaches $K_{IC}$ this denominator term approaches zero. The crack growth rate of equation (5.4) then increases without bound.

The last term in the numerator introduces the threshold behavior, i.e. as $\Delta K$ approaches $\Delta K_{th}$ from above, the numerator, and hence the rate of growth, goes to zero.

An examination of equation (5.4) reveals that the dimensional units of the constant $C$ will differ for different values of the exponents. An alternative form of equation (5.4) which exhibits the features of the complete curve but eliminates the changes in dimensional units of $C$ can be developed. To illustrate this, consider an equation of the form

$$\frac{da}{dN}=C'\frac{(1-R)^m\left[\dfrac{\Delta K-\Delta K_{th}}{K_0}\right]^p\left(\dfrac{\Delta K}{K_0}\right)^n}{(1-R)^q\left[1-\dfrac{K_{max}}{K_{IC}}\right]^q}.$$

(5.6)

By this use of dimensionless ratios, the primary features of the growth rate curve are retained, but the constant $C$ always has the same units as $da/dN$. The constant $K_0$ could, moreover, be taken as the critical stress intensity factor.

An examination of equation (5.4) reveals that the rate of crack growth is a function of the applied loading and the geometry of the cracked body. The functions for $K$ can be determined either by use of available tabulations or by analysis. Equation (5.4) can then be integrated to determine the crack growth history. Numerical integration will often be required. The process can, however, be illustrated by simplifying the growth rate equation. If the Paris equation is considered, then

$$\Delta a=C\left(K_{max}-K_{min}\right)^n\Delta N.$$

(5.7)

For a tensioned sheet with a central crack for which the crack length is very small compared with the sheet width

$$K_1 = \sigma \sqrt{\pi a}, \tag{5.8}$$

where $\sigma$ is the applied tensile stress. The crack length is $2a$. It follows from equations (5.7) and (5.8) that

$$C \int_0^{N_f} \frac{dN}{\left(\sigma_{max} - \sigma_{min}\right)^n} = \int_{a_0}^{a_f} (\pi a)^{-\frac{n}{2}} \, da, \tag{5.9}$$

where $a_0$ is the initial crack length and $a_f$ amd $N_f$ are the final crack length and number of cycles, respectively. If $(\sigma_{max} - \sigma_{min})$ is held constant and $n = 3$, equation (5.9) can be integrated to give

$$N_f = \frac{2\left(\sigma_{max} - \sigma_{min}\right)^3}{\pi^{3/2} C} \left( a_0^{-\frac{1}{2}} - a_f^{-\frac{1}{2}} \right). \tag{5.10}$$

When the maximum stress intensity factor becomes equal to $K_{IC}$, the crack length $a_f = a_c$. The value of $a_c$ can be obtained by use of equation (5.8). That is,

$$a_C = \frac{1}{\pi} \left( \frac{K_{IC}}{\sigma_{max}} \right)^2. \tag{5.11}$$

Using this result in equation (5.10) gives the critical number of cycles, $N_c$.

The use of equation (5.11) implies that the crack is extending in mode I. In thin sheets a crack which initially grows in mode I is sometimes observed to undergo a transitional behavior. The fracture surface rotates from being on a plane normal to the direction of loading to a plane which mades an angle of about 45 degrees relative to the loading direction. After the rotation, the propagation of the crack continues in the original direction, but under a mixture of modes I and III. Equation (5.10) cannot, therefore, be used after such a transition occurs. The problem of fatigue crack growth under mixed mode conditions is discussed in section 5.6.

Since it was discovered that $\Delta K$ could be utilized as a correlation parameter for fatigue crack growth, a number of specimens have been used. The specimens described in Appendix B provide a few examples of possible choices. Often the selection of a specific type is based on size limitations of the material stock to be tested.

Monitoring crack growth has been accomplished by use of a number of measurement techniques. These include the use of optical instruments, compliance gauges and electrical potential difference devices, and acoustic emission and ultrasonic techniques. Good descriptions of these methods are contained in a collection of articles edited by Beevers (1980).

The concepts of linear elastic fracture mechanics and the use of the stress intensity factor are limited to cases in which small scale yielding is developed at the crack tip. In thick specimens a state of plane strain is developed at the crack tip. The stress state in the interior then approaches a triaxial tension and yielding is suppressed. This suppression is not present in a cracked, thin sheet for which plane stress conditions are developed. The plastic zone size for the plane strain condition can be shown to be about one-third of that for the plane stress case. For monotonic loading the approximate radius of the plastic zone size, $\rho_p$, for plane strain is

$$\rho_p = \frac{1}{3\pi} \left( \frac{K_I}{\sigma_y} \right)^2 , \tag{5.12}$$

where $\sigma_y$ is the yield strength. The corresponding radius for plane stress is about three times larger. Some limitations on the use of $K_I$ for estimating plastic zone sizes are discussed in section 2.4.4.

The plastic zone produced at a crack tip undergoing cyclic loading differs from that produced by monotonic loading. By assuming that reversed plastic flow can occur during unloading, that superposition of loading and unloading states is permissible, and that the fracture surfaces do not come in contact with one another, it can be shown that the cyclic plastic zone can be smaller than that for monotonic loading to the same maximum stress (Rice, 1967). Microhardness test measurements made by Bathias and Pelloux (1973) have provided evidence which supports this conclusion.

By applying the above rationalization, the cyclic plastic zone can be estimated by substituting $\Delta K$ for $K$ and multiplying $\sigma_y$ by a factor of 2.

Thus, for plane strain

$$\rho_{cyc} = \frac{1}{3\pi} \left( \frac{\Delta K}{2\sigma_y} \right)^2 . \qquad (5.13)$$

The value of $\sigma_y$ used here should be for the cyclically stabilized state (section 4.3).

Equation (5.13) can be solved for $\Delta K$ and rewritten as

$$\Delta K = 2\sigma_y \sqrt{3\pi \rho_{cyc}} . \qquad (5.14)$$

A comparison of Figs 3.2 and 5.1 now indicates that the abscissa parameters differ by the quantity $2\sigma_y\sqrt{3\pi}$. This equivalence holds, of course, only if equation (5.13) is valid. The possible use of $\rho_{cyc}$ as a correlation parameter when the use of $\Delta K$ is not valid, is discussed in section 5.4.

If the net stress on the cross-section adjacent to the crack approaches the yield stress, small scale yielding no longer prevails, and the stress intensity factor does not characterize the crack tip state. For the case of elastic – plastic fracture mechanics Rice's invariant contour integral, $J$, can, however, be used to characterize the crack tip state.

The use of a range of $J$, $\Delta J$, has been proposed as a correlation parameter for fatigue crack growth for elastic – plastic load cycling (Dowling, 1977; Starkey and Skelton, 1982). The $J$ integral was, however, proposed for use with monotonic loading. For elastic – plastic fatigue the stress – strain path for loading is nonlinear and that for unloading is linear elastic. Thus, although elastic – plastic fatigue crack data appear to be correlated by this application of $\Delta J$, support for its use is based on an empirical rationalization rather than theoretical arguments.

Another restriction of the use of $\Delta J$ may arise in its application to the problem of variable amplitude loading. The generation of $da/dN$ versus $\Delta J$ diagrams is based on a procedure which requires the availability of stabilized, cyclic stress – strain curves. For variable amplitude loading, however, transient states rather than stable states are developed in the enclave at the crack tip. The importance of variable amplitude loading effects is discussed in section 5.3.

## 5.2.2 The threshold phenomenon

For most components in service the capability of surviving a very large number of loading cycles prior to failure is desirable. It follows that the rate of crack growth during at least the initial period of growth should be small. This in turn indicates a need for analyses which predict crack growth rates for values of $\Delta K$ which are initially, at least, slightly greater than $\Delta K_{th}$, i.e. below the region in which the Paris equation is valid.

Although it is reasonable to expect that the crack growth rate in the threshold region is a function of $\Delta K$, it clearly is not described by the Paris equation. One approach which has been suggested is to replace $\Delta K$ in the Paris equation by an effective $\Delta K$. This has focused attention on a partitioning of threshold behavior mechanisms. According to this concept, the crack growth equation may be written in the form (Beevers and Carlson, 1986)

$$\frac{da}{dN} = C\left(\Delta K - \Delta K_{th}\right)^{n}.$$ (5.15)

The intrinsic and extrinsic parts of $\Delta K_{th}$ may be written as

$$\Delta K_{th} = in\,\Delta K_{th} + ex\,\Delta K_{th}.$$ (5.16)

In section 3.2 it is observed that the operative mechanisms in the near threshold region are distinct from those which prevail in region II (Fig. 3.3). The departure of the rate of growth curve in region I from a backward extension of the Paris equation line should, therefore, not be unexpected. Whereas crack extension occurs by the formation of striations in region II, a faceted, cleavage crack surface is observed in region I. The features of the resulting crack surfaces in region I are microstructure dependent, and this dependence is manifested in the intrinsic component of $\Delta K_{th}$.

The extrinsic component of $\Delta K_{th}$ has been associated with crack face features which deviate from those assumed for the analytical model of a crack. Evidence for the existence of a variety of these features has been obtained, and they all result in obstructions to complete crack closure during the unloading phase of a load cycle. The obstructions include the presence of debris, the formation of a plastic wake, the development of crack face roughness, the formation of an oxide, and the presence of a fluid.

The obstructing effect of debris was reported by Christensen (1963), who conducted fatigue crack growth tests on 2024-T3 clad aluminum alloy sheet and reported that 'increases in fatigue life for the crack growth phase of test panels possessing trapped metal fragments are attributed to reductions in the resultant range of working stress at the tip of the growing crack'. The closure obstruction mechanism was discovered by placing cellulose tape on the sheet faces to prevent debris from being extruded from between the crack faces. Without the tape, debris were not trapped and fatigue lives were shorter.

Subsequently, Elber (1971) conducted experiments which indicated that as a fatigue crack advances, the plastic deformation developed at the crack tip recedes from the tip and forms a wake which acts as an obstruction to complete closure. Elber's experiments were conducted on a sheet material, so nominally plane stress conditions could be expected to have been developed.

In continuing research Purushothaman and Tien (1975), Walker and Beevers (1979), Halliday and Beevers (1981), and Ritchie and Suresh (1981) discovered that crack surface roughness could also produce obstruction to closure. This form of obstruction was observed primarily for plane strain conditions. The deviation of the roughened fatigue fracture surfaces from an idealized planar crack is nicely illustrated in Fig. 5.2 (a) and (b) for an unloaded specimen of the aluminum alloy 2024-T351 (Halliday, 1994). In Figure 5.2 (a) jogs appeared to be of the order of the grain size. In Figure 5.2 (b), where the magnification has been increased by a factor of ten, the jogs appear to be of the order of 0.1 the grain size. Thus, the apparent size of the asperity steps is dependent on the scale of observation. The irreversible distortion which accompanies crack growth has prevented complete closure and has resulted in discrete points of closure contact. It may also be seen that both vertical and horizontal forces could be introduced at the points of contact. Both mode I and mode II conditions could, therefore, be introduced by these local contacts. The consequences of this type of mixed mode loading are discussed in section 5.6.

An environment in which the crack faces are subject to oxidation has also been reported to lead to closure obstruction in steels, aluminum alloys and copper (Paris, *et al.*, 1972; Ritchie, Suresh and Moss, 1980; Suresh, Zamiski and Ritchie, 1981; Liaw, *et al.*, 1982; Vasudevan and Suresh, 1982).

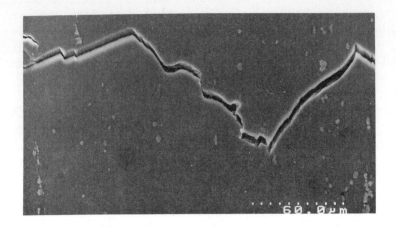

(a)

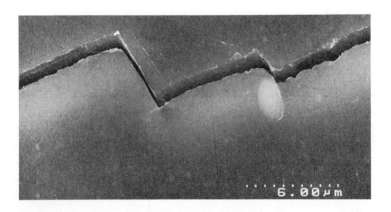

(b)

**Fig. 5.2**  Fatigue crack profiles in aluminum alloy 2024-T351 (Halliday, 1994)

In a fundamental paper Budiansky and Hutchinson (1978) demonstrated that the material in the crack tip plastic zone could, as a crack advanced under cyclic loading, result in a wake which becomes an obstruction to complete closure. They assumed small-scale yielding conditions and used a Dugdale – Barenblatt model as a basis for their analysis. Their results confirmed the expectation that the presence of such an obstruction results in an effective stress intensity factor range which is less than that determined from the externally applied loading. Since they did not, however, introduce a criterion for crack advance, their results are not suitable for crack growth predictions. Also they observed that 'the Dugdale model is most appropriate for plane stress problems, whereas plane strain conditions are generally more relevant to fatigue crack growth. For this reason, we do not propose to explore the implications of the present model to great depth'.

Newman (1981) has provided a good description of the features of a modified Dugdale strip model of which various forms have been used by a number of researchers. The strip models proceed by continuously computing the thickness of a perfectly plastic layer and introducing a crack extension criterion to enable crack growth to be determined. To extend the Dugdale model restriction from plane stress to plane strain conditions, a constraint factor is introduced. The modified Dugdale model has been incorporated into a predictive fatigue crack code by Newman (1992a). Predicted crack growth rates and fatigue lives have been shown to be in good agreement with experimental results.

McClung (1992) has conducted comprehensive elastic – plastic, finite element analyses of fatigue crack growth and has presented results which exhibit both the closure obstruction and the growth rate features which have been experimentally observed. The effects of stress level, stress ratio, strain hardening, elastic constraint and variable amplitude loading have been examined in these studies.

Blom (1993a) has presented a review in which a Dugdale strip model and finite element results are compared with data from fatigue crack growth tests. He concludes that the use of a strip model analysis is adequate for engineering purposes. Heuler and Schütz (1985) have suggested that a safety factor of at least 2 should probably be applied for crack growth predictions to allow for the uncertainties encountered in applications to aircraft analyses. Some of the topics which have an influence on these uncertainties are discussed in sections 5.3 through 5.6.

It has been suggested (Knott, 1986) that the effective distances of obstructions behind a crack tip may range from less than 0.1 mm for roughness and oxide closure tip to values greater than 1 mm for plastic wakes. It should also be noted that an obstruction produced by a tensile overload may be operative in preventing closure until it is replaced by a new, closer effective obstruction (Carlson, Kardomateas and Bates, 1991).

The common feature of all of the obstruction mechanisms is that they reduce the effective value of the stress intensity factor range. That is, because of closure obstruction

$$\Delta K_{eff} < (K_{max} - K_{min}).$$

The details of this behavior can be illustrated by the use of a simple model which places a single obstruction at an effective distance behind the crack tip (Beevers, et al., 1984).

Although contact during closure can be expected to occur at discrete obstructions along and across crack face surfaces, it is assumed in the model that these can be replaced by a single, equivalent asperity acting at an effective distance from the crack tip. As formulated, the model may be analyzed as a plane strain problem in which closure obstruction is represented by opposing line forces acting on the crack faces. The cracked body model is shown in Fig. 5.3 (a),where each crack face force is $P$. External loading is represented by two symmetrically applied forces of magnitude $Q$.

The total stress intensity factor is the sum of a 'local' stress intensity factor due to the crack face forces and a 'global' stress intensity factor due to externally applied forces. By superposition, the total stress intensity factor is

$$K_I = K_I(\text{local}) + K_I(\text{global}). \tag{5.17}$$

Geometric details of the model are shown in Fig. 5.3 (b), where $C$ is the effective distance of the obstruction from the crack tip. The local stress intensity factor is then

$$K_I(\text{local}) = \left(\frac{2}{\pi C}\right)^{\frac{1}{2}} \frac{P}{B}, \tag{5.18}$$

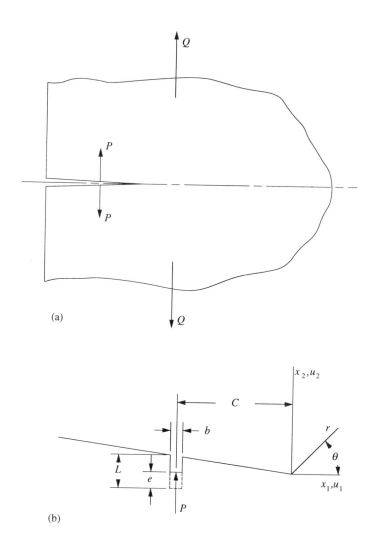

(a)

(b)

**Fig. 5.3** (a) External and crack face loading and (b) discrete asperity model

where $B$ is the thickness of the cracked body. This result is for an infinite body with a semi-infinite crack, but here the distance $C$ is very much smaller than the typical body width, so the equation should apply.

The dimension $L$ represents the magnitude of the interference produced by the asperity obstruction. Until the external loads are sufficient to eliminate contact between the upper asperity and the lower asperity, there will be a compression of an amount $e$. The effective width of the asperity is $b$.

The global stress intensity factor depends on the body geometry and the loading, and it can be either determined by reference to handbooks or by a finite element analysis.

The vertical displacement of any point on the upper crack face (at $r$ and at $\theta = \pi$) may be written as

$$U_2(r,\pi) = U_2(\text{global}) + U_2(\text{local}).\tag{5.19}$$

By use of the stress intensity factors for these two cases, we can write the displacement at $r = C$, $\theta = \pi$ as

$$U_2(C,\pi) = \frac{2}{\mu}\left(\frac{C}{2\pi}\right)^{\frac{1}{2}}(1-v)K_{\mathrm{I}}(\text{global}) + \frac{2}{\pi\mu}(1-v)\frac{P}{B}.\tag{5.20}$$

The compression of the asperity is

$$e = \frac{PL}{EbB}.\tag{5.21}$$

The displacement of the crack face may then be written as

$$U_2(C,\pi) = (L-e) = \frac{2}{\pi\mu}(1-v)\frac{P}{B} + \frac{2}{\mu}\left(\frac{C}{2\pi}\right)^{\frac{1}{2}}(1-v)K_{\mathrm{I}}(\text{global}).\tag{5.22}$$

The quantities $e$ and $P$ can be determined by use of equations (5.21) and (5.22). The quantities $C$, $L$ and $b$ are related to the microstructural details for a given material.

The stress intensity factor with zero external load and the stress intensity factor for the case in which the external load reduces the local contact load to zero can be determined as special cases of the above equations. These are referred to as the closure and opening stress intensity factors, respectively.

For the case of zero external load, $K_I(\text{global}) = 0$, and the use of equations. 5.18, 5.21 and 5.22 gives

$$K_I = K_I(\text{local}) = \left(\frac{2}{\pi C}\right)^{\frac{1}{2}} \left|\frac{1}{Eb} + \frac{2}{\pi \mu L}(1-v)\right|^{-1} \tag{5.23}$$

where $E$ is Young's modulus, $\mu$ is the rigidity modulus and $v$ is Poisson's ratio.

In the case in which the external load is just sufficient to remove the asperity pressure, $P = 0$, and $e = 0$. Then equation (5.22) yields the result

$$K_I = K_I(\text{global}) = \frac{L\mu}{2(1-v)} \left(\frac{2\pi}{C}\right)^{\frac{1}{2}} \tag{5.24}$$

The features of the model can be illustrated by reference to a diagram of the total stress intensity factor versus the external load in Fig. 5.4. With no closure obstruction, the loading path cycles along the line OA. If, during unloading from point A, closure obstruction is encountered at point B, the model for a single asperity results in a straight-line load path that moves downwards and to the left of B, but above line OB. For two asperities, it can be shown that two straight-line segments will be developed (Carlson and Beevers, 1984).

The reduction in the effective range of the stress intensity factor can be illustrated by considering a test in which the maximum load is $Q_{max}$ and the minimum load is zero. The effective range of the stress intensity factor is then given by the distance between points C and D. This gives a value which is significantly less than a value based solely on the global loading. This latter value is given by the distance between points C and O.

The point D represents the 'closure' value of $K_I$. The point B represents the 'opening' value. The distance between C and F is sometimes taken as a measure of the effective stress intensity factor.

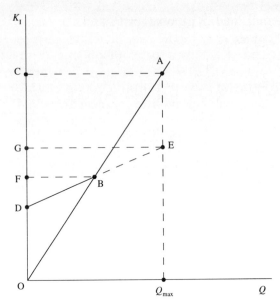

**Fig. 5.4**   Effect of closure obstruction on stress intensity factor

Another feature of the model can be used to rationalize the difference in crack growth rates for active and inert atmospheres. A welding mechanism has been used as an explantion for experimental results obtained by Kendall and Knott (1984). They found that crack growth rates for a low carbon steel were smaller in a vacuum than in air. They attributed the difference to greater reversibility of a slip mechanism which leads to rebonding or welding of slip surfaces in a vacuum. In air, contamination of the slipped surfaces inhibits slip reversibility and results in an increment of crack growth. An alternative mechanism may be identified by noting that if, during closure, asperity contacts became welded to one another, the model solution for $Q_{min} = 0$ would then cycle along the extended line $D - B - E$ instead of along the line $D - B - A$. From B to E the asperities would be under a tensile load. The first path represents asperity welding and the corresponding effective $\Delta K_I$ ranges from D to G. The second path describes a case in which no welding occurs and the effective $\Delta K_I$ ranges from D to C. The effective $\Delta K_I$ for the inert atmosphere case is smaller than that for the active atmosphere case, so the crack growth rate for the former case would be expected to be smaller than that for the latter case.

The discrete asperity model is most clearly identified with surface roughness. All of the closure mechanisms, however, possess the same obstruction characteristics. The asperity model can, therefore, also be used to illustrate the consequences of other forms of obstruction to closure.

It is clear from the preceding discussion that obstruction to closure plays an important role in fatigue crack growth in the near threshold region. If, during unloading, contact between the crack faces occurs before the minimum load is reached, obstruction is a factor. If, however, the minimum load occurs before such contact, closure obstruction is not a factor. The conditions of loading which separate these two loading states have been nicely demarcated by use of a special test procedure developed by Hertzberg *et al* . (1992). In these experiments the value of $K_{max}$ is maintained constant, and the value of $K_{min}$ is gradually increased. This results in a decreasing $\Delta K$ and an increasing stress ratio value, $R$. Ultimately, closure becomes negligible, so the boundary between closure and nonclosure is established. The use of this technique was also recommended for use in determining the effects of closure in small cracks (section 5.4).

## 5.3 VARIABLE AMPLITUDE LOADING

### 5.3.1 Introduction

The discussion of fatigue crack growth in the preceding sections was concerned with nominally steady state loading conditions. Thus, although $\Delta K$ changes as a crack grows under fixed loading conditions (e.g. $P_{max}$ and $P_{min}$ constant), the changes are continuous, not abrupt. For components in some machines, this type of loading is reasonably representative of the actual conditions which are encountered. Components in nonstationary structural systems such as aircraft are, however, exposed to histories in which the loading magnitudes change abruptly. These loading histories are often described as being random and in fatigue considerations they are characterized as variable amplitude loading conditions.

The effects of variable amplitude loading on fatigue crack behavior introduce a number of special problems. These include the determination of load spectra, and the development of methods in which laboratory data obtained under steady state conditions can be used to analyze and predict crack growth under variable amplitude loading. Issues involving additional topics which should be examined include mixed mode loading, small crack behavior and environment. How these affect fatigue crack growth in metals, nonmetals and composites is considered in the sections which follow.

## 5.3.2  Loading spectra

The variable amplitude loading history or spectrum for a structural system is required input for the determination of fatigue crack growth. The development of loading spectra for aircraft has of necessity been the objective of intensive studies, so a review of procedures which have been adopted forms a basis for an introduction to the subject. It is clear that a capability for analyzing fatigue crack growth is a prerequisite for the development of a damage tolerance design procedure for a new aircraft. Also, however, it can be a part of a program in which aging aircraft which were originally designed using a safe life strategy are being re-examined for damage tolerance.

One approach which can be utilized is to use strain gauge instrumentation which monitors outputs at critical locations during representative ground – air – ground missions. This is both time consuming and costly, and can normally be justified only if the results can be used as input for a fleet of aircraft which have similar missions. An advantage of its use is that it can form the basis for a cycle by cycle analysis of fatigue crack growth.

In the course of developing a preliminary design, which involves choices of materials and alternative designs, decisions based on damage tolerance considerations are required. For this type of problem standards of loading which can be used to evaluate the options are available. The loading programs cited in section 2.2 for various aircraft provide an example of spectra which can be used for this type of evaluation. A description of the characteristics and the use of these spectra is contained in the referenced article by ten Have (1989).

Although the load spectra which are available represent loading conditions which can be encountered in flight, they cannot be used directly in the determination of fatigue crack growth. There is, however, more than one procedure for processing the load spectra into a form in which individual cycles can be identified. Emphasis in the more generally accepted procedures is based on the identification of load ranges rather than peak values. The most generally accepted method of determining equivalent cyclic spectra is the 'rainflow' method.

Elements of the 'rainflow' method can be visualized by reference to Fig. 5.5. Flow begins at each reversal in the load record. Flow stops if it encounters flow from above (segments $C - B'$, $F - E'$ and $H - G'$). It also stops if it reaches a point which is opposite a peak for which the absolute value is greater than that from which the flow started (compare points B and D, E and G, G and A'). The cycle ranges for this example are: $A - D - A'$, $B - C - B'$, $E - F - E'$, $G - H - G'$. A review of these ranges indicates that all segments of the original record are included as parts of the four ranges. A concise review of the evolution of cycle counting procedures is presented in the text by Bannantine, Comer and Handrock (1990).

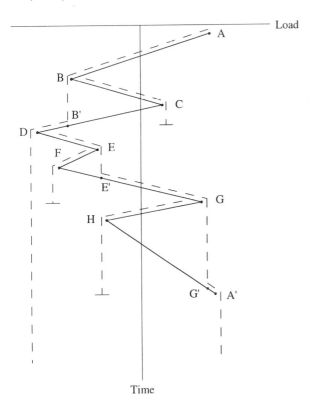

**Fig. 5.5** Rainflow counting

### 5.3.3 Load interactions

The accuracy of predictions of fatigue crack growth which result from random loading depends on the degree to which the cycle counting procedure which is used provides a valid representation of the effect on growth of the actual loading record. It also depends upon the extent to which the analytical procedure used to determine the crack growth history incorporates the effects of load interactions during variable amplitude loading. If the equation for crack growth rate is simply integrated on a cycle by cycle basis, the effect of load history will not be included. This would be analogous to the Miner – Palmgren rule, i.e. the change in crack length during a cycle is independent of previous loading cycles. This does not, for example, recognize that closure obstruction can be expected to vary with variations in the maximum load.

Correlations between crack growth test data and analytical predictions for random loading can differ depending on how the comparisons are made. Partl and Schijve (1990) have observed that the apparent accuracy of a correlation can depend upon the type of comparison which is utilized. Predictions can, for example, be based on a cycle by cycle examination. This requires a fractographic examination of the type performed by Partl and Schijve. The number of cycles to failure or the crack length at the end of a given number of cycles can also be used as a basis for comparison.

An attempt to evaluate some of the methods of analysis which have been proposed has been made by designing and implementing a large 'round-robin' program (Chang and Hudson, 1981). In this program random load spectra for four fighter mission types and one transport mission were applied to 2219-T851 aluminum alloy center-cracked-tension specimens. Different load factors were used to produce a total of 13 test cases. The experimental data were processed using six different methods of analysis. Of these, two incorporated schemes for including load interaction effects. In summarizing the results Chang (1981) noted that the average of the ratios of predicted to test cycles to failure for 76 analyses was 1.28, with a standard deviation of 0.45. It should, however, be noted that the method which had the smallest ratio value (1.07) had prediction ratios which ranged from 2.52 to 0.64. This suggests that the recommendation (section 5.2.2) of Heuler and Schütz (1985) of using a safety factor of 2 for predictions is well founded.

The results of the round-robin program tended to give non-conservative predictions, and it was suggested that the sparcity of data in the near threshold region might have contributed to this trend.  An examination of Fig. 5.6, which shows  the crack growth rate in the near threshold region, reveals that the rate is very sensitive to  small changes in $\Delta K$, i.e. large changes in rate can occur with small changes in $\Delta K$. The detailed features of the equations used to represent crack growth rate (equation 5.4) differ. These differences and scatter in the experimental data used to generate crack growth equations can be expected to contribute uncertainty to predictions of crack growth history.

The uncertainties cited above have been compounded by the observation that very short cracks have been found to grow below the threshold value of $\Delta K$ (this behavior is discussed in section 5.4).  The recognition of the potential problems associated with representations of crack growth rate in the near threshold region has resulted in a number of proposed remedies.  The most simple of these is the suggestion that the rate in this region be described by an extension of the Paris equation (Owen, Bucci and Kegarise, 1989).  This is shown as a dashed line in Fig. 5.6.  Although this can be expected to result in conservative predictions, it may provide a margin of safety which is part of the prediction method as opposed to a factor of safety which is applied at the end of the analysis.

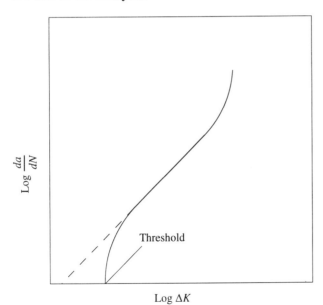

**Fig. 5.6**  Near threshold features of a *da/dN* versus $\Delta K$ diagram

### 5.3.4 Tensile overloads

The difficulties encountered in developing crack growth prediction models can be recognized by restricting attention to the transient effects which have been observed after a single overload. The mechanisms that have been proposed to explain the transient retardation after a single tensile overload include residual stress (Schijve, 1960), crack deflection (Lankford and Davidson, 1982), crack closure (Elber, 1971), strain hardening (Jones, 1973) and plastic blunting/ resharpening (Christensen, 1959). Of these mechanisms, two have been the subject of the most intensive examination and application. These are the effects of residual stress and crack closure. Willenborg, Engle and Wood (1971) and Wheeler (1972) proposed models which include the effect of residual stress. Versions of these models have been adopted for use in fatigue crack growth programs.

The transient retardation which follows a tensile overload has been the subject of a number of experimental investigations. To clearly focus on the transient response, some investigators have conducted experiments in which a constant stress intensity range, $\Delta K$, is maintained except for a single, tensile overload.

Ward-Close and Ritchie (1988) conducted tensile overload experiments on compact tension specimens which were machined from IMI 550 titanium alloy plate. All tests were performed under predominantly plane strain conditions. The test results were obtained on specimens with a fine grained $\alpha/\beta$ microstructure. In their experiments a stress intensity factor range of $\Delta K = 15$ MPa$\sqrt{m}$ with $R = 0.1$ was maintained constant. Single 100% and 150% tensile overloads for which $K_{OL} = 2K_{max}$ and $2.5K_{max}$, respectively, were applied.

McEvily and Yang (1990) conducted similar experiments on the aluminum alloy 6061-T6. For their tests $\Delta K = 8.0$ MPa$\sqrt{m}$ and $R = 0.05$. An overload of $\Delta K = 16.0$ MPa$\sqrt{m}$ with $K_{min}$ unchanged was used.

Since the Willenborg model has received considerable attention, a review of its features can contribute to an understanding of its applicability to the results obtained by Ward-Close and Ritchie and by McEvily and Yang.

A residual stress intensity factor, $K_R$, is introduced to account for the compressive residual stress state introduced by a tensile overload. Since $K_R$ is subtracted from both $K_{max}$ and $K_{min}$, $\Delta K$ is not changed. The ratio of $(K_{min} - K_R)$ to $(K_{max} - K_R)$ is, however, decreased, so the effective $R$

ratio is decreased. This would be expected to lead to a decrease in the crack growth rate.

By definition

$$K_R = \phi \left[ K_{OL} \left( 1 - \frac{a}{\rho_{OL}} \right) - K_{max} \right] \tag{5.25}$$

where

$$\phi = \left[ 1 - (K_{th} / K_{max}) \right] (S - 1)^{-1}$$

with $a$ being the crack extension after the overload, $\rho_{OL}$ the plastic zone size of equation (5.12), $K_{th}$ the threshold stress intensity factor and $S$ the shut-off ratio.

A measure of the duration of retardation is the distance that a crack advances during the transient behavior following an overload. This has been called the 'delay distance'. To provide the required decrease in crack growth rate, $R$ must be decreased. This requires that $K_R$ must initially be positive. Since it decreases to zero as the crack advances through the overload plastic zone, predicted delay distances can be determined by setting $K_R = 0$. These predicted delay distances have been compared with experimentally measured delay distances for the data obtained by Ward-Close and Ritchie and by McEvily and Yang (Carlson, Kardomateas and Bates 1991) and the results are presented in Table 5.1.

Table 5.1 Delay distances

| Material | Measured delay | Willenborg delay (plane strain) (mm) |
|---|---|---|
| Titanium alloy | 0.88 | 0.06 |
| Aluminum alloy | 0.35 | 0.18 |

The differences between the measured and Willenborg delay distances indicate that the development of a compressive residual enclave cannot be the primary operative mechanism for the observed retardation. It is of interest to note that other forms of the square bracketed term in equation (5.25) could be used to exhibit the retardation behavior, and they would have different delay distances. This is illustrated as an example in problem 5.10.

For the tests conducted by Ward-Close and Ritchie (1988) and McEvily and Yang (1990), obstruction to closure was observed, and the opening values of the stress intensity factor, $K_{op}$, were determined for crack extension after overloads were applied. The data of Ward-Close and Ritchie are presented in Fig. 5.7. The data in Fig. 5.7 (a) are for an overload of 100%. Those of Fig 5.7 (b) are for an overload of 150%. As plotted, the overloads occurred at the origin. In the interval between the origin and the first point, a transient crack growth accleration or delayed retardation was observed. This behavior had been observed previously by Nowack *et al.* (1979) and by Paris and Hermann (1982).

This immediate, post-overload behavior was attributed to a crack tip stretching which reduced closure contact pressures between previously formed obstructions and, by effectively increasing the stress intensity factor range, resulted in a brief crack growth acceleration period. Ultimately, the crack extended through the stretched ligament and formed an asperity which resulted in the $K_{op}$ data points shown in Fig. 5.7. It should be noted that the immediate post-overload transient cannot be rationalized by use of residual stress models of the Willenborg type.

Experimental results obtained by Buck, Thompson and Rehbein (1988) and by Ward-Close and Ritchie (1988) have indicated that the primary crack face contacts which result from surface roughness induced obstruction are immediately adjacent to the crack tip. The discrete asperities model provides a relatively simple means of restricting attention to the near crack tip region of closure contact. This does not mean that the segments of the crack face more distant from the crack tip are neglected. It simply means that the more distant asperities are a part of the past crack advance history which does not influence current behavior.

Equation (5.24) represents the value of $K$ at which, during loading, the contact force goes to zero, and it has been identified as the opening stress intensity, $K_{op}$. which is then

$$K_{op} = \frac{L\mu}{2(1-v)}\left(\frac{2\pi}{C}\right)^{\frac{1}{2}} \tag{5.26}$$

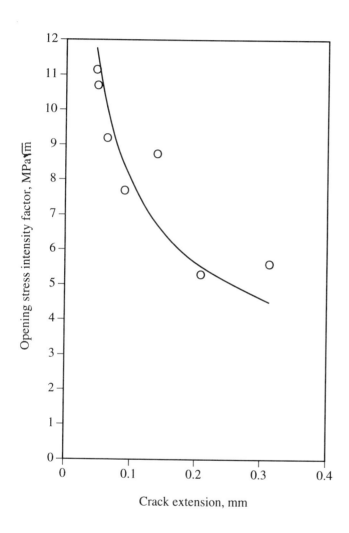

**Fig. 5.7(a)**   Opening stress intensity factor versus crack extension for a 100% tensile overload (from Ward-Close and Ritchie, 1988)

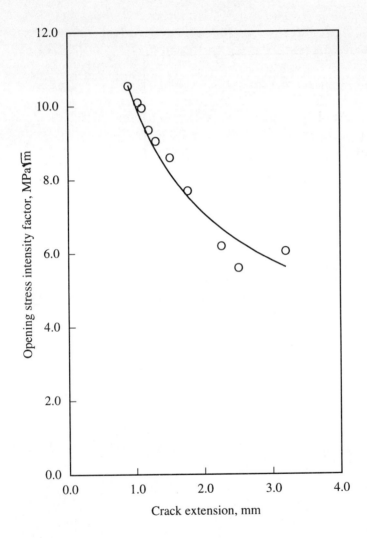

**Fig. 5.7(b)**  Opening stress intensity factor versus crack extension for a 150% tensile overload (Ward-Close and Ritchie, 1988)

The variable $C$ in equation (5.26) represents the distance of the asperity from the crack tip. As the crack extends, $C$ increases as the asperity recedes from the crack tip. For a rigidity modulus $\mu = 41.2$ GPa, $v = 0.3$ and an asperity misfit value of $L = 1.1\mu m$, equation (5.26) yields the curve shown in Fig. 5.7 (a) for a tensile overload of 100%. An obstruction misfit value $L = 4.3$ $\mu m$ gives the curve of Fig. 5.7 (b) for an overload of 150%. These curves describe how $K_{op}$ decreases as the distance increases between the tip of the growing crack and the asperity produced by the overload. An examination of these plots reveals that the data points are well correlated by the use of equation (5.26).

In the experiments analyzed, retardation did not begin until values of $C$ of about 0.04 mm and 0.65 mm after the overloads were achieved. This immediate, post-overload behavior has been attributed to crack tip stretching which has already been discussed.

Obstruction to closure has been observed to occur in different forms. In sheets, the obstruction is formed at the specimen faces at the free ends of the crack tip front. For dominantly plane strain conditions, obstructions in the form of surface roughness on the interior crack surfaces can be developed.

The experiments of McEvily and Yang (1990) were designed to examine the form of closure obstruction after a tensile overload. They presented evidence which indicated that even though crack closure obstruction prior to an overload was of the plane strain form, obstruction after an overload occurred at the specimen faces, i.e. under plane stress conditions. Data points for $K_{op}$ obtained by McEvily and Yang (1990) on the aluminum alloy 6061-T6 are presented in Fig. 5.8. The trend of the data points is similar to that observed by Ward-Close and Ritchie. To examine the form of the obstruction resulting from the overload McEvily and Yang repeated their experiment, but after the overload, they removed surface layers from the specimen by machining. Upon resumption of testing, no retardation transient was observed. They concluded that the obstruction to closure after the overloads occurs at the side faces of the specimens. This would appear to resemble the 'pop-in' behavior which is observed in fracture toughness tests. The magnitude of the overload may be expected to have an influence on this behavior.

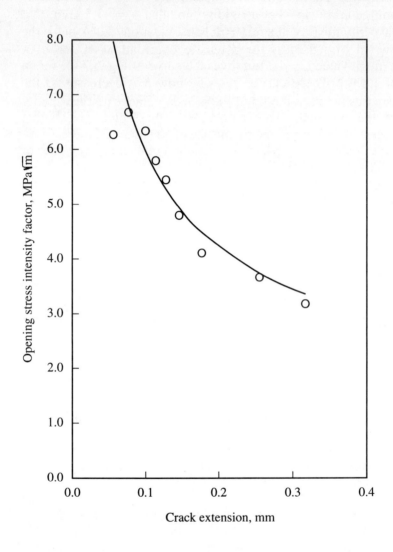

**Fig. 5.8**  Opening stress intensity factor versus crack extension (McEvily and Yang, 1990)

The curve in Fig. 5.8 was obtained by applying equation (5.26) to the data of McEvily and Yang for the aluminum alloy 6061-T6. The values used to generate the curve were $\mu = 26.5$ GPa, $v = 0.3$ and $L = 1.26$ µm. Although the obstruction to closure in this test occurred at the surfaces, the correlation between data points and the use of equation (5.26) is good.

The discrete asperity model is two-dimensional and therefore may be considered to be more applicable to through-the-thickness closure. The correlation with the McEvily and Yang (1990) data, however, suggests that $L$ can be considered the 'effective' value of obstruction which props the crack open.

The importance of closure obstruction on crack growth behavior in the near threshold region is generally accepted and the results presented in Figs 5.7 and 5.8 provide evidence for this view. The results also provide support for the use of analyses which incorporate the effect of obstruction to closure. This includes predictive codes which are based on the use of a modified Dugdale model (Newman, 1981).

### 5.3.5  Compressive overloads

A number of computer codes have been developed for determining fatigue crack growth histories for variable-amplitude loading. A feature of many of the service load spectra is the presence of intermittent compressive excursions. Since it has seemed reasonable to believe that no contribution to crack growth is developed during a compressive excursion, it has been a common analytical practice to exclude the affects of negative loads (Bucci, 1981). Thus, for a negative $R$ cycle the effective stress intensity factor range is set equal to the maximum stress intensity factor.

The effect of compressive load excursions has been the subject of a number of experimental programs and a review of the results indicates that it is incorrect to assume that compressive cycles do not contribute to fatigue crack growth. The use of this assumption can, in fact, lead to non-conservative predictions.

Compressive loading excursions have been investigated in tests on both smooth bar specimens and cracked specimens. Zaiken and Ritchie (1985) conducted fatigue crack growth tests on compact tensile specimens of a cast I/M 7150 aluminum alloy. They found that crack growth could occur for loading levels below the threshold stress intensity range after the application of large compression overloads. They attributed this behavior to a flattening of previously formed roughness asperities which reduced the effective stress intensity range.

Pompetzki, Topper and DuQuesnay (1990) have conducted extensive investigations of the effects of overloads on fatigue damage in smooth bars of the aluminum alloy 2024-T351 and in SAE 1045 steel. They applied fully reversed ($R = -1$), constant-amplitude load cycles to specimens with circular cross-sections to obtain uniaxial stress amplitude versus cycles to complete failure to establish the baseline behavior.  In additional experiments they repeated the previous tests, but applied intermittent compressive overloads.  Three sets of tests were conducted, in which the number of cycles between the intermittent overloads differed.  All of the intermittent overload test curves were below the baseline curve for stress versus cycles to failure.  The level of the curves decreased with decreasing cycles between overloads.  Both the aluminum alloy and the steel exhibited this behavior.  The authors suggested that the observed behavior resulted from a decrease in the crack closure opening stress which resulted in an increase in the crack growth driving force.  The reduction in opening stress was attributed to a decrease in the height of the crack tip wake by yielding during compressive overloads.

Yu, Topper and Au (1984) have presented test results on the aluminum alloy 2024-T351 in the near threshold region. They found that crack growth curves shifted to lower threshold values and greater crack growth rates as the minimum stress became more compressive. Yu , Topper and Au (1984) examined the effects of single, intermittent compressive  excursions in tests which were otherwise loaded under fixed positive $R$ values.  The tests were conducted on center-crack plate specimens.  The number of cycles between the intermittently applied loads was varied from test to test.  The number of cycles between the compressive load cycles ranged from 1 to infinity for six values. Introduction of the compressive cycles substantially shifted the growth rate curves, with the highest rates being developed for one excursion per cycle (all negative $R$ cycles), and the lowest for infinite cycles between excursions (no negative $R$ cycles).  The shift behavior was observed even when only one compressive cycle was applied between 1000 positive $R$ cycles.

The variation in response of three alloys was investigated by Kemper, Weiss and Stickler (1989).  The alloys – age-hardened aluminum 2024, pure copper and the powder metallurgy aluminum alloy IN-905XL – had different microstructures, deformation behavior and mechanical properties, and the authors demonstrated that the responses to negative $R$ loading differed significantly. They concluded from their experimental results that the recommendation of using only the tensile portion of cyclic loading was valid only when obstruction to closure was absent. Surface roughness measurements indicated that this applied only to the ultra-fine-grained, mechanically alloyed Al IN-905XL for which the fracture surface was very flat. It also appeared to be valid for the soft copper under very high compressive loadings. It was not valid for the Al 2024, for which very high compressive loads were not able to reduce the level of closure obstruction.

Most of the experimental results that have been reported have dealt with the effect of compressive excursions on the rate of crack growth versus the range of the stress intensity factor plots. Results have been presented that focus directly on crack growth behavior (Carlson *et al.* 1993). Data were presented on three alloys: Wasapaloy, the aluminum powder metallurgy alloy IN-9052, and the bearing steel M50 NiL.

Crack growth was measured for tests in which constant values of maximum stress and minimum stress with positive $R$ were applied. The tests were then interrupted, and although the maximum stress was maintained, the minimum stress was reduced to a negative, compressive value so that $R$ was negative. After an interval, the initial loading was resumed. Fig. 5.9 illustrates the type of loading that was applied.

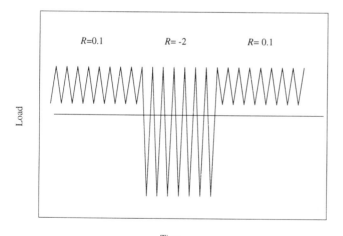

**Fig. 5.9**  Loading sequences

An examination of the data for Waspaloy in Fig. 5.10 reveals that the rate of crack growth (the slope) is discontinuous at each change in loading condition. For these tests $\sigma_{max}$ = 300 MN $m^{-2}$ and the initial crack length $a_0$ = 1.20 mm. Also, it is clear that the rate of growth for $R$ = $-$ 2 loading is substantially greater than that for $R$ = 0.1. It also appears that the final slope for the initial $R$ = 0.1 phase is slightly greater than the initial slope for the final $R$= 0.1 loading. This suggests that the interposed $R$ = $-$ 2 loading may have introduced a transient retardation behavior upon resumption of the $R$ = 0.1 loading.

The most obvious feature of Fig. 5.10 is the abrupt increase in the rate of crack growth at the loading transition from $R$ = 0.1 to $R$ = $-$ 2. If the growth rate were unaffected by the compressive portion for the ratio value of $R$ = $-$ 2 loading, there would be no change in the slope of the growth curve. This behavior may, at least in part, be attributed to a difference in the effective stress intensity factor range for the two loading conditions. This can be shown by reference to features of a discrete-asperity model for closure obstruction which is illustrated in Fig. 5.11 as a plot of the variation of the mode I stress intensity factor with external load. With no closure obstruction, the loading path cycles along the line OA. If, during unloading from point A, closure obstruction is encountered at point B, the model for a single asperity results in a straight-line load path that moves downwards and to the left of B, but above line OB. For two asperities, it can be shown that two straight-line segments will be developed. For a number of asperities an increasing number of contacts will occur during unloading, and the curve to the left of point B represents this behavior. If the heights of the asperities are inelastically reduced, however, it may be anticipated that unloading would drop below the elastic solution, as shown by the dashed curve.

Consider two loading conditions: $R$ = 0 and $R$ = $-$ 2. For cyclic loading between O and $Q_{max}$ ($R$ = 0) the range of the effective stress intensity factor would be measured on the $K$ axis from point D to point C in Fig. 5.11. $K$ is greater than zero because of closure obstruction. For $R$ = $-$ 2 loading, the range of the effective stress intensity factor would be measured from point F to point C. Since the magnitude of the distance from F to C is greater than that from D to C, it would follow that the rate of crack growth for $R$ = $-$ 2 would be greater than that for $R$ = 0. The difference depends upon the distance from F to D.

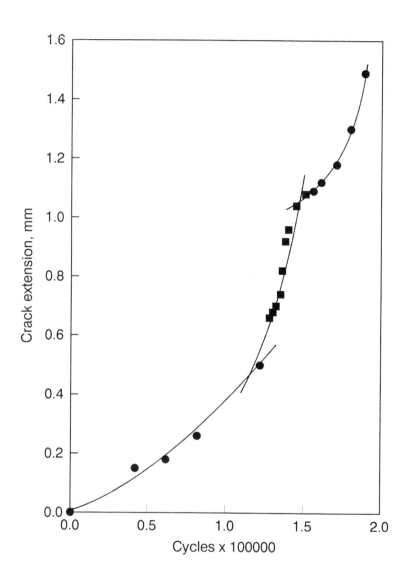

**Fig. 5.10**  Compressive excursion data for Waspaloy (Carlson et al., 1993)

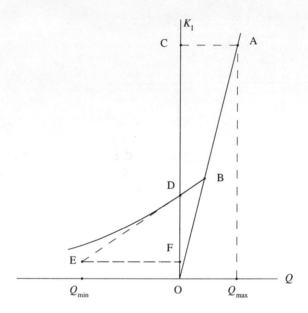

**Fig. 5.11** Effect of closure obstruction on stress intensity factor

Tests on IN-9052 and M50 NiL exhibited the same type of abrupt difference in the rate of crack growth between the positive and negative $R$ values. The initial, transient retardation behavior exhibited by the Waspaloy was observed for M50 NiL, but not for IN-9052. It has been suggested (Carlson and Kardomateas, 1994) that the transient behavior occurs during an interval in which the nominally steady state condition of one $R$ ratio value changes to that of the subsequent $R$ ratio. Thus, both the closure obstruction features and the cyclic plastic zone sizes undergo an accommodation to the new loading conditions. The 'steady state' cyclic plastic zone size for $R = -2$ in Fig. 5.10 would be larger than that for $R = 0.1$, and the fatigue crack must traverse the $R = -2$ zone before the $R = 0.1$ zone is established. Note that cyclic hardening or softening may also play a role.

The features of the experimental results that have been reported indicate that the observed behavior is primarily associated with a closure effect. The degree to which closure obstruction affects the observed compressive overload behavior depends upon the fracture surface misfit and the hardness of the metal. These, as noted by Kemper, Weiss and Stickler (1989), depend upon microstructural features.

Both Kemper, Weiss and Stickler (1989) and Tack and Beevers (1990) observed a saturation effect in which increases in compressive load beyond a certain level do not result in additional increases in crack growth rate. Reduction of asperity heights could be subject to a limiting inelastic deformation as the asperity cross-sectional areas are increased under increasing compressive loading. Strain-hardening effects could also be expected to contribute to the saturation behavior. It may further be observed that the type of discrete contact associated with roughness asperities could be expected to provide for more 'efficient' crushing than a continuous wake layer.

It is of interest to note that the limitation on the crushability of the closure obstructions may be expected to produce incomplete crack closure even for very large compressive loads. Tack and Beevers (1990) conducted corner fatigue crack tests with negative $R$ values on three bearing steels. They obtained crack tip micrographs which indicated that complete crack tip closure did not occur for the bearing steels. Micrographs for the Rolls Royce steel, grade RBD are shown in Fig. 5.12. In Fig. 5.12 (a) the tensile load was +20 kN. In Fig. 5.12 (b), when a compressive load of $-50$ kN was applied to give an $R = -2.5$, there was an open gap adjacent to the crack tip.

An inelastic asperities model has been developed (Kardomateas and Carlson, 1995) and both the effects of the crushing of closure obstructions and saturation with increasing compression are exhibited in applications of the model.

Under variable amplitude or random loading, 'steady state' conditions are not developed and the consequences can be inferred from the compressive and tensile overload experiments. They indicate that the rate of crack growth during a given cycle can depend upon loading which occurred during a previous cycle. The resulting crack growth can, therefore, be viewed as a succession of intervals during which transient behavior occurs. Or, more succinctly, the crack growth is history dependent.

Efforts to include the effects of history have been based primarily on the development of methods for determining an effective range of stress intensity factor which takes into account the load interactions which occur during variable amplitude loading. For examples, see Chang and Hudson (1981).

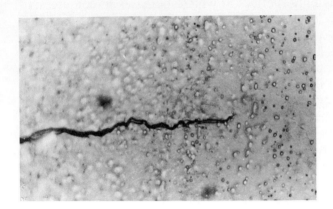

(a)

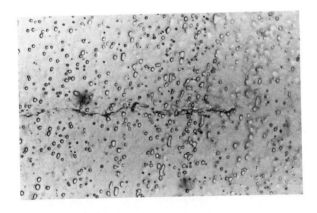

(b)

**Fig. 5.12** Crack closure for RBD bearing steel for (a) a tensile load of +20 kN (1000x) and for (b) a compressive load −50 kN (1000x) (Tack and Beevers, 1990)

## 5.4 SMALL CRACKS

### 5.4.1 Introduction

In section 5.2.1 it was noted that the range of the stress intensity factor could be used to estimate the size of the cyclic plastic zone at the tip of a fatigue crack. Equation (5.12) provides the means of obtaining this estimate. In a discussion of the limitations of small scale yielding Rice (1967) predicted that if the use of the range of stress intensity factor underestimates the cyclic plastic zone size (equation 5.13) for short cracks, they should grow faster than long cracks for the same calculated range of stress intensity factor values. This can be illustrated by considering an edge crack in a wide plate for which

$$\Delta K = 1.1(\Delta\sigma)(\pi a)^{1/2}, \tag{5.27}$$

where $\Delta\sigma$ is the applied stress range and a is the crack length. For the same range of stress intensity factor values a 'small' crack of length 0.30 mm must have a $\Delta\sigma$ which is almost 6 times larger than that for a 'long' crack of length 10 mm. If the stress intensity factor is to be used to analyze small fatigue crack data, the limitations of small scale yielding need to be checked. That is, to what extent is the use of $\Delta K$ valid in the correlation of small fatigue crack data?

Widespread interest in the behavior of small fatigue crack growth was initiated after experimental results published by Pearson (1975) became well known. Pearson discovered that small cracks could grow under values of $\Delta K$ which were less than the threshold value. The subject, which is sometimes described as the 'anomalous' behavior of small fatigue cracks, has attracted the attention of many research investigators, and has been the topic of symposia edited by Miller and de los Rios (1986), Ritchie and Lankford (1986) and Larsen and Allison (1992).

It should be noted that both 'short' and 'small' have been used in the literature to describe the fatigue crack behavior of interest in this section. An Appendix to ASTM Test Method E647 (1993), however, makes a distinction between these two descriptions. It defines a small crack as one for which all crack dimensions are small compared to a relevant parameter such as a microstructural feature. For a surface crack the dimensions are the depth and the surface length. A crack is designated as being short when only one of the dimensions is small, e.g. the depth dimension of a through-crack in a plate can be small, but the length of the crack front equals the plate thickness which is not small.

The dependency of fatigue behavior on crack length can be demonstrated by use of a Kitagawa diagram (Kitagawa and Takahashi, 1976). For the diagram in Fig. 5.13, $R = 0$ and the lower bound for fatigue failure for very small cracks in region I is given by the value of the endurance limit stress. In region III linear elastic fracture mechanics applies and the lower bound is given by the equation

$$\Delta\sigma_{th} = \frac{\Delta K_{th}}{Y\sqrt{\pi a}} \, , \tag{5.28}$$

where the subscript 'th' denotes threshold value and $Y$ is a coefficient for specimen geometry and loading.

The behavior of small cracks in region II is not described by extensions of either of the straight lines of regions I and III. Rather, observed data points fall below these extensions as indicated by the solid curve.

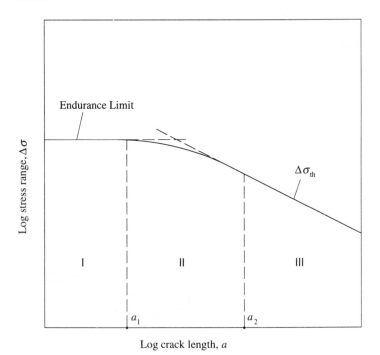

**Fig. 5.13** The Kitagawa diagram

The values of the crack lengths $a_1$ and $a_2$ are dependent on microstructural features and can, therefore, vary widely. Taylor and Knott (1981) found that for a number of metals $a_2$ was approximately equal to $10d$, where $d$ is a microstructural unit size, such as the grain size.

Although the prediction of Rice (1967) regarding the behavior of small fatigue cracks was based on analytical results, the primary impetus for subsequent research has evolved from experimental observations. The types of specimens used and the methods of crack measurement are, therefore, of special importance in a description of the behavior that has been observed. Since the cracks and the crack growth increments are of the order of microstructural unit sizes, such features as grain size, texture, precipitates and inclusions also influence the nature of crack initiation and growth processes.

### 5.4.2 Experimental results

Small cracks can be initiated in specimens used to obtain stress versus cycles to failure diagrams of the type shown in Figs 4.1 and 4.2. The locations of these cracks cannot be predicted, however, and cracks often occur at multiple sites. To provide control over the crack initiation site, a commonly used practice is to introduce either a small stress concentration or a small flaw. Versions of the specimen shown in Fig. 5.14 (a) have been used frequently (Blom *et al.* 1986; Larsen, Jira and Ravichandran, 1992; Newman, 1992b). Crack initiation in these specimens is confined to the bottom of a semi-circular notch of radius $R$. Although cracks are sometimes initiated at one of the corners, the surface crack depicted at the interior location is more common.

Surface cracks in plates and cylinders, and through-thickness edge and corner cracks in bars with rectangular cross-sections can be initiated by the introduction of the small flaws indicated in Fig. 5.14 (b), (c) and (d). These can be introduced by the use of electrical discharge machining or by thin cutoff wheels (Gangloff *et al.* 1992; Pickard, Brown and Hicks, 1983). Detailed recommendations for the preparation of specimens and descriptions of crack growth measurement techniques are presented in an appendix to ASTM Test Method E647 (1993). More detailed discussions of crack measurement methods are given in the proceedings of ASTM STP 1149 (eds J. Larsen and J.E. Allison, 1992).

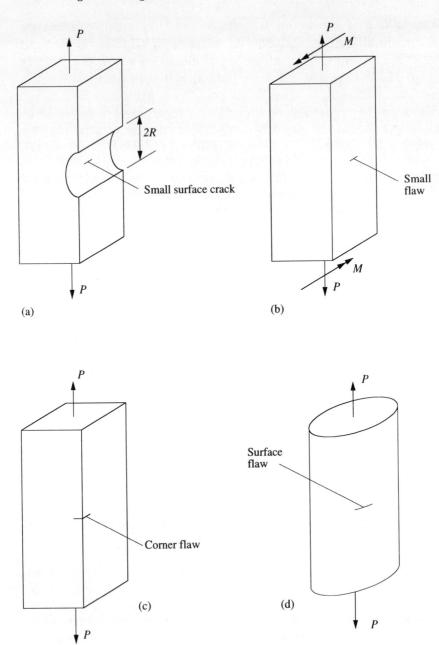

**Fig. 5.14** Small crack specimens

The methods included are replication (Swain, 1992), photomicroscopy (Larsen, Jira and Ravichandran, 1992), scanning electron microscopy (Davidson, 1992), an interferometric strain/displacement gauge method (Sharpe, Jira and Larsen, 1992), electrical potential drop (Gangloff *et al*. 1992) and ultrasonics (Resch and Nelson, 1992).

Microstructural influences on small crack growth have been identified by several investigators. Blom *et al* . (1986) found that crack nucleation emerged from fractured intermetallic particles in the aluminum alloy 2024 and along particle – matrix interfaces and in slipbands within a grain in the aluminum alloy 7475.

Swain (1992) has presented results for single edge notch tension specimens and reported that small surface cracks initiated in pits from which inclusions had been removed during surface preparation in a 4340 steel and in the aluminum alloy 7075-T6. Swain (1992) also described tests in which small cracks were initiated at more than one site. The introduction of criteria for accepting or rejecting test results for multiple site cracking was, therefore, considered necessary.

The importance of grain size and shape has been revealed by observations that small cracks can move quickly through grains, but are retarded when they reach a grain boundary. This behavior has been observed in the aluminum alloy 2219-T851 by Morris (1979), in low carbon steels (Tanaka, Nakai and Yamashita, 1981), in an aluminum alloy (Lankford, 1982), in titanium alloys (Wagner *et al*. 1986) and in nickel base alloys (Sheldon *et al*. 1981). Although the grain boundary barrier may appear to be of secondary importance for polycrystalline metals, it should be noted that the number of grains along the front of a small corner crack is not large. For a grain size of 100 µm the number of grains along the front of a crack which is 0.25 mm deep is about 4. For a through crack in a 6 mm thick plate the corresponding number of grains would be about 60. Similar considerations for small surface cracks in an aluminum alloy have been presented by Bolingbroke and King (1986).

The texture developed in material stock depends on the manufacturing process used. For the plate material tested by Blom *et al*. (1986) the grain dimensions for the aluminum alloy 2024 were 120 µm in the longitudinal direction, 65 µm in the transverse direction and 25 µm in the short transverse direction. The propagation of a small crack in this alloy could, therefore, be expected to depend on the direction of growth.

The leading edges of surface cracks have often been observed to exhibit semi-circular contours. Ravichandran and Larsen (1992) have, however, presented evidence which indicates that the geometry of the crack front deviates from the semi-circular pattern in highly textured metals. Blom *et al.* (1986) found that as a small surface crack in the aluminum alloy 2024 grew, the ratio of the depth to the half surface length changed from 0.9 to 0.6, i.e. the crack shape changed from semi-circular to semi-elliptical. Texture, grain size and the distribution of inclusions and precipitates can, therefore, be expected to effect the initiation and growth of small fatigue cracks.

Kitagawa diagrams of the type shown in Fig. 5.13 describe a boundary below which fatigue failure does not occur. Stress and crack length coordinates above this boundary can be expected to lead to crack growth and ultimately to fracture. It is of interest at this point to examine an extension of the diagram to include an upper bound. Remembering that $R = 0$, we can locate the value of the ultimate strength on the stress axis. The value of the yield strength, $\sigma_y$, is shown between the values of the endurance limit and the ultimate strength, $\sigma_u$. Also, above the lower bound in region III, it should be possible to determine the stress corresponding to the critical stress intensity factor. Thus, since $\Delta\sigma = \sigma_{max} = \sigma_c$ for this loading condition

$$\Delta\sigma = \sigma_c = \frac{K_{IC}}{Y\sqrt{\pi a_c}}. \tag{5.29}$$

Using these results, the Kitagawa diagram can be extended to include the upper boundary shown in Fig. 5.15. Note that the upper boundary is analogous to the separation of mechanisms which are operative for elastic and inelastic column buckling.

The upper bound, correspondingly, has a segment on the right which represents fracture which can be described by elements of linear elastic fracture mechanics (LEFM). The segment on the left represents the unstable condition for elastic – plastic fracture mechanics (EPFM). For initial values of stress and crack length between the upper and lower boundaries a point will move, under constant stress, to the right until it reaches the upper boundary; i.e., the crack will extend until a critical condition is attained. The paths $A_o - A_c$ and $B_o - B_c$ represent two small crack growth histories.

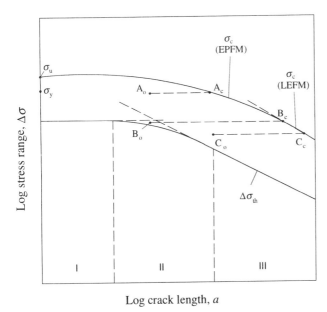

**Fig. 5.15**   An extended Kitagawa diagram

It should be noted that the diagram of Fig. 5.15 provides no information on the origin of the cracks depicted. This is a basic feature of fracture mechanics. The crack originating at point $A_o$ clearly is a small crack. The primary focus, however, in much of the research on the 'anomalous' behavior of small cracks is on stress-crack length values corresponding to point $B_o$, i.e. below the extended lines for $\sigma_e$ and $\sigma_{th}$. This point is below the threshold, and reported data on small cracks indicate that crack growth can occur for the indicated initial values of stress and crack length.

Conditions of the type represented by point $C_o$ can be analyzed by the use of $\Delta K$ as the driving force for crack growth. The crack growth behavior for the conditions of point $A_o$ would, however, be expected to require an elastic – plastic analysis. The conditions for point $B_o$, which is below the extended threshold line, are initially not characterized by the use of the parameter $\Delta K$.

The sizes of the cyclic plastic zones at the crack tips can be expected to differ for these examples. Point $A_o$ can be expected to develop large scale yielding. Initially, small scale yielding would be developed for the point $C_o$ condition. The scale of yielding for point $B_o$ would be intermediate between those for points $A_o$ and $C_o$. This conclusion regarding point $B_0$ provides an implicit basis for efforts directed toward the development of methods which have been proposed for the use of modified values of stress intensity factor range.

### 5.4.3 Analytical considerations

Larsson and Carlsson (1973) have shown that for long cracks the size of the plastic zone at the crack tip can be significantly underestimated by the use of the stress intensity factor. They found, however, that the size of the zone computed by an elastic – plastic finite element analysis can be closely estimated by including the second, constant term of the Williams series (section 2.4.4). Kardomateas *et al.* (1993) have presented analyses which suggest that the addition of the second term can have a similar effect for small cracks. These results indicate that the use of modified or effective stress intensity factors may be acceptable within an intermediate range between large and small scale yielding.

Often, when small crack data are presented, the stress intensity factors have been determined for the given crack length and shape, and clusters of points are plotted on *da/dN* versus $\Delta K$ diagrams. As would be expected from the location of small crack data points in Fig. 5.15, the clusters are to the left of the curve for *da/dN* versus $\Delta K$ in the near threshold region. These results reinforce the conclusion that $\Delta K$ may not be a proper correlation parameter for small fatigue cracks. It should also be noted that the lack of continuous curves in the representations of actual data is a consequence of the fact that the growth rates are initially cyclic in character. As noted previously, periods of growth and retardation are observed as crack fronts move through grains and encounter barriers such as grain boundaries. Plots of crack growth versus cycles of loading are, therefore, more revealing about the nature of the growth of small cracks.

Two types of small crack growth behavior can be illustrated by the representations shown in Fig. 5.16. The dashed curve on the left represents growth under a decreasing rate which ultimately leads to arrested growth. The dashed curve on the right indicates an initial period of decreasing rate followed by an increasing rate as the crack grows to a length for which the behavior is governed by the concepts of LEFM. Although the two decreasing rate behaviors are illustrated in Fig. 5.16, the operative mechanisms are not characterized by the parameter $\Delta K$. As used here, the parameter merely indicates how the growth rate changes for this particular arrangement of the stress and crack length variables. To provide another perspective of crack behavior, the points $A_0$, $B_0$ and $C_0$ of Fig. 5.15 have been positioned on the graph of Fig. 5.16. $C_0$ represents long crack behavior. Both $A_0$ and $B_0$ are small cracks. The stress level for point $A_0$ is, as noted earlier, in the elastic – plastic regime, and it could represent, for example, a small crack at the root of a notch.

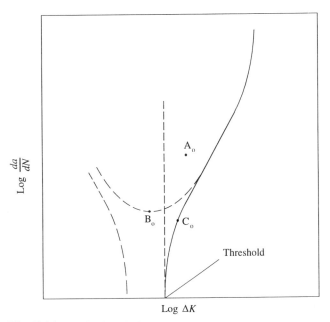

**Fig. 5.16**  Typical depiction of long and small crack behavior

Cracks which are smaller than can be detected by nondestructive inspection techniques (less than 0.5 mm) may ultimately become long cracks, so states below the threshold cannot be ignored. A variety of approaches have been proposed to address this problem. Perhaps the most simple is that proposed by Owen, Bucci and Kegarise (1989). This proposal, which is depicted in Fig. 5.6, suggests that the *da/dN* versus $\Delta K$ curve be extended to the left from the Paris region. Although this does not directly address the small crack issue, it does provide a simple procedure for adjusting the crack growth rate equation to compensate for the small crack behavior.

Blom *et al.* (1986) conducted experiments in which the load, after crack nucleation, was successively decreased and increased to determine the levels at which crack growth was arrested and resumed.  By this means, they were able to establish a range over which a transition from small to long crack behavior occurred. They then constructed a line which was parallel to the line for $\Delta K_{th}$ (Fig. 5.13) and passed through the point on the boundary at which the transition from region I to region II behavior occurs.  The line so constructed was defined as the eff$\Delta K_{th}$ and they suggested that it be used in LEFM crack growth laws.  This process can result in a shifting of the lower part of the growth rate curve to the left.  The  value of the eff$\Delta K_{th}$ would have to be experimentally determined for each material of interest.

Blom *et al.*. (1986) partially attribute the transition behavior to a change in the magnitude of the closure obstruction.  They, as well as others, suggest that closure obstruction is either absent or nearly absent for very small cracks, and that it increases with increasing crack length.

Edwards and Newman (1990) have proposed the use of a modified or effective $\Delta K$ which is dependent on the value of applied stress at which crack opening occurs.  They reason that if closure obstruction is smaller for small cracks than for long cracks, a decrease in the opening stress, $S_0$, should result in an increase in the effective stress intensity factor $\Delta K_{eff}$.  The relation

$$\Delta K_{eff} = \frac{(S_{max} - S_0)\Delta K}{(1-R)S_{max}}$$

(5.30)

was proposed for use in a predictive code based on a modified Dugdale model (Newman, 1981).  These analyses have been used to correlate data from tests on long and small cracks and for both constant and variable amplitude loading conditions (Edwards and Newman, 1990).

### 5.4.4  Elastic – plastic crack growth

When the applied stress is near or exceeds the yield strength, the use of modified or effective values of $\Delta K$ no longer has a rational basis for acceptance.  The states developed under this level of loading may be considered to correspond to the 'low cycle' fatigue behavior discussed in section 4.3.

Correlations of elastic – plastic crack growth data have often been based on a modified version of the invariant contour integral $J$ of Rice (1968). For LEFM $J$ is related to $K$ by the equation (2.49). The use of this equation is restricted to applications governed by linear elasticity. The use of $J$, however, is valid for cracked bodies with nonlinear stress-strain relations. It has, therefore, been applied to elastic – plastic problems in fracture mechanics.

The use of $J$ in elastic – plastic fracture mechanics has primarily been applied to problems involving monotonically increasing tensile load. Dowling (1977), recognizing the fact that in fatigue the range of loading provides the driving force for crack extension, proposed the use of a parameter defined as $\Delta J$.

For a given cracked body, $J$ can be evaluated by the use of Rice's invariant contour integral. Techniques for determining $J$ from a knowledge of the stress-strain curve and the geometry and loading of the cracked body have also been developed. Shih and Hutchinson (1976) have, for example, developed an equation for $J$ for a center cracked specimen under tensile plane stress loading. For a plastic exponential hardening material they found that

$$J_{pl} = 2\pi \, W a f(s) \tag{5.31}$$

where $W$ is the area under the stress – strain curve up to the value of stress for the given loading, $2a$ is the crack length and $f(s)$ is a function of the strain hardening exponent. Using a stabilized, cyclic stress – strain curve, and the suggestion of Shih and Hutchinson (1976) that elastic and plastic solutions for $J$ can be added, Dowling (1977) proposed the following equation for $\Delta J$:

$$\Delta J = C_1 \Delta \, W_e \, a + C_2 \Delta \, W_p \, a \tag{5.32}$$

The $\Delta W_e$ and $\Delta W_p$ are the areas defined as shown in Fig. 5.17. Cyclic unloading for the stabilized hysteresis loop is shown as a dashed curve. The coefficients $C_1$ and $C_2$ are crack geometry constants. For a surface crack in a circular cylinder, Dowling estimated that $C_1 = 3.2$ and $C_2 = 5.0$. It should be noted that the experiments which Dowling conducted were for an $R = -1$. The original origin, denoted as O, was, therefore, shifted as shown for the calculations.

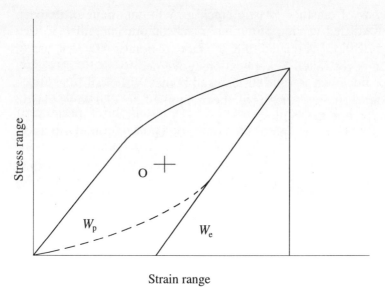

Fatigue crack growth correlations presented by Dowling (1977) for A533B steel were sufficiently encouraging to create an interest in the use of Δ*J*, and a number of investigators conducted studies on its application.   Starkey and Skelton (1982), for example, used Δ*J* to analyze data for  small surface fatigue cracks in high strength ferritic steels, and they concluded from their results that Δ*J* was effective in correlating short crack growth rates.   They also suggested that the hysteresis loop should be bounded in the lower half at the closure stress to exclude the area formed while the crack was closed. Their computation of $W_e$ and $W_p$ was then based on an upper portion of the hysteresis loop.

McClung and Sehitoglu (1988) have presented results of small fatigue crack growth experiments on a 1026 steel.  Cracks were initiated at the root of very small notches and fully reversed strain ranges were applied.   In addition to correlating data by use of Δ*J*, they also made measurements and performed analyses on the use of the range of crack tip opening displacement, ΔCTOD, as a correlation parameter.  In addition to demonstrating that both Δ*J* and ΔCTOD provided good correlations of their data, they also examined the effects of closure and opening loads. They performed experiments in which blocks of constant strain cycles were interrupted by significantly larger strain excursions. During the large strain excursions, the crack tip was stretched so that during subsequent cycling at the smaller strain range, the crack remained open during the entire cycle for a transient period. After the stretched region receded a sufficient distance from the growing crack tip, closure contact prior to the minimum load was resumed. The intervals during which the crack was open, however, resulted in brief periods of accelerated growth. It should be noted that this overload behavior appears to be analogous to the initial, delayed retardation behavior observed for long cracks after a tensile overload (section 5.3.2).

Methods for determining $J$ from experiments on cracked bodies have been proposed by Rice, Paris and Merkle (1973) and by Clarke *et al.* (1976). A method for the experimental determination of $J$ from a hysteresis loop has also been proposed (Merkle and Corten, 1974). These methods make use of load – displacement records from tests on the cracked body of interest.

Hutchinson (1983) has discussed conditions which must be satisfied to insure that $J$ provides a valid characterization of nonlinear fracture. These include monotonic and proportional loading. The latter requires that the ratios of the stress components at each point remain constant during loading. No rigorous proof has been provided that these conditions are satisfied for the type of loading applied in fatigue crack tests. It may also be observed that the application of $\Delta J$ has made use of stable hysteresis loops. In problems involving variable amplitude loading such stability is not likely to be achieved in the cyclic plastic zone. The use of $\Delta J$ to compute crack growth histories in predictive codes could, therefore, introduce uncertainties which may be difficult to evaluate.

The $\Delta$CTOD measurements made by McClung and Sehitoglu (1988) are closely related to the crack tip cyclic plastic zone. Nisitani, Goto and Kawagoishi (1990) have proposed that a measure of the cyclic plastic zone should be used as the driving force for the rate of crack growth. Recognition of this relationship served as a motivation for using the cyclic plastic zone as a correlation parameter in Fig. 3.3. The relation which Nisitani, Goto and Kawagoishi proposed is

$$\frac{da}{dN} = D \left( \frac{\Delta\sigma}{\sigma_y} \right)^n a, \tag{5.33}$$

where $D$ and $m$ are material constants. Although they used this law to correlate small fatigue crack data, it does not provide a continuous transition from small to long cracks. Also, it does not conveniently incorporate, as $K$ does, a description of specimen geometry and loading.

Newman (1992) has proposed the use of a cyclic plastic zone corrected range of effective stress intensity factor which he defines as

$$\Delta \overline{K}_{eff} = (S_{max} - S_0')\sqrt{\pi d} \quad F\left(\frac{d}{w}, \frac{d}{r}, ...\right) \tag{5.34}$$

where $S_{max}$ is the maximum applied stress, $S_0'$ is the crack opening stress and $F$ is a boundary correction factor which depends on geometric ratios of $d$ to specimen width $w$, notch or hole radius $r$, etc. The quantity $d$ is a fictious crack length which is adjusted to include part of the cyclic plastic zone size. The suggested equation for $d$ is

$$d = c + \gamma \omega \tag{5.35}$$

where $c$ is the actual crack length and $\gamma$ is a factor determined from an analysis of the cyclic $J$ integral (Newman, 1992b). An estimate of $\omega$ is

$$\omega = \tfrac{1}{4}\rho\left(1 - \frac{S_0'}{S_{max}}\right)^2 \tag{5.36}$$

where $\rho$ is the Dugdale plastic zone size.

Newman (1992b) has successfully used equation (5.34) to predict small crack growth for variable amplitude and spectrum loading tests conducted in several different laboratories. Focusing on the size of the cyclic plastic zone appears, therefore, to provide a basis for analyzing the elastic – plastic growth of cracks.

## 5.5  STRESS CONCENTRATION EFFECTS

Components in service often have geometrical features, such as holes, fillets or notches, which act as stress raisers. Thus, although the nominally applied stress may be small, the stress concentration site can experience large stresses.

If the stress at the stress concentration site is less than the yield strength, LEFM conditions exist and the primary problem is the determination of the effect of the stress gradient on the value of the stress intensity factor. Some insight into this problem may be gained by reference to the four cracked specimens in Fig. 5.18. All are infinite plates. Two, (a) and (c), have elliptical holes with a radius of $\rho$ at the ends of the major axis. The crack lengths for (a) and (b), and for (c) and (d) are the same. All are exposed to the same cyclic tensile stress, $\sigma_\infty(t)$, with $R = 0$.

By virtue of the difference in local stresses, crack initiation can be expected to occur after many more cycles for case (b) than for case (a). Also, for small crack lengths, $l$, the rate of growth for case (a) will initially be greater than that for case (b).

After the cracks have grown to the lengths indicated in Fig. 5.18(c) and (d), however, the driving force for the two cases will have become essentially equal. The crack emanating from the elliptical hole will have extended beyond the influence of the notch affected region.

Since it was initially required that the level of stress adjacent to the root of the notch be elastic, the applied stress level, $\sigma_\infty(t)$, must be relatively small. For a circular hole, for example, it would have to be less than $0.333\sigma_y$. One might, therefore, speculate that once the crack emanating from the ellipse extends beyond the region of influence of the stress concentration, it may enter a phase in which the growth rate decreases. Experimental results reported by Frost and Dugdale (1957) and by Frost (1960) have indicated that crack growth can, in fact, even become arrested. Frost, Marsh and Pook (1974) have characterized the arrest behavior in terms of the applied stress range and the stress concentration factor, $K_t$. The general conclusion of these studies is that arrest occurs above a critical value of $K_t$ and below a threshold value of stress range. Since arrested growth has been associated with 'sharp' notches, a steep stress gradient may be necessary.

Newman (1971) has proposed an equation for the stress intensity factor of the form

$$K_I = 1.12\, FK_t\, \sigma^\infty \sqrt{\pi l} \tag{5.37}$$

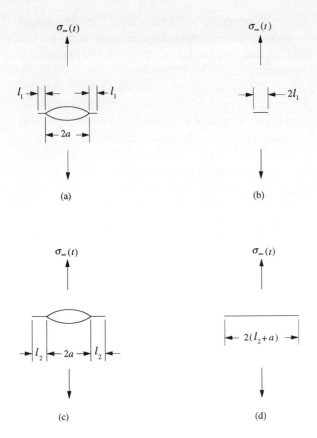

**Fig. 5.18** Comparisons of center cracks and cracks at the ends of an elliptical hole

Lukás and Klesnil (1978) have developed a relationship for $F$ of the form

$$F = \left[1 + 4.5\left(\frac{l}{\rho}\right)\right]^{\frac{1}{2}}$$ 
(5.38)

where $l \leq 1.5\rho$.

When the stress at the stress concentration site reaches the yield strength, further increases in the maximum applied load will produce a stress redistribution, and a plastic enclave will be formed at the ends of the major axis. Under cyclic loading, the local loading of an element at a stress concentration site will essentially be exposed to low cycle fatigue. Crack nucleation can, therefore, be anticipated to occur in a relatively small number of cycles.

The size of the plastic enclave can, depending on the level of loading, range from a locally contained zone to a fully plastic state along the horizontal axis. For the fully plastic case cyclic loading would result in an unacceptably small number of cycles to failure. Primary interest in applications is, therefore, on loading conditions which result in contained plastic flow. After a crack advances beyond the initial plastic zone adjacent to the notch, a period of retardation is sometimes observed. This can be followed by either an increasing rate of crack growth or arrest, depending upon the level of applied loading.

The methods described for use in the analysis of small cracks when LEFM conditions do not apply can also be considered for these problems. These include the uses of $\Delta J$ and equation (5.34) which are discussed in section 5.4.4. It should also be noted that Moftakhar and Glinka (1992) have proposed a method for the elastic – plastic analysis of notched bodies. They have shown that upper and lower bounds on notch tip strain energy can provide a basis for determining bounds for the stress and strain components under proportional loading. They suggest that knowledge of the stress and strain distributions can be used in fatigue life prediction models based on cyclic strains at a notch tip.

## 5.6 MIXED MODE STATES

In previous sections the emphasis has been on mode I fatigue crack growth. Components in machines and structures, however, can develop cracks which are not exposed exclusively to mode I conditions. This can be a consequence of the local geometry and the type of loading, and it can result in a crack plane which is curved rather than flat. For a three dimensional crack more than one mode can be present, and the ratios of the stress intensity factors can, in fact, vary along the crack front. This is illustrated for a surface crack in Fig. 2.8. A recognition of the importance of mixed mode states is indicated by the fact that several conferences have been concerned with technical issues which have been identified (Sih and Theocaris, 1981; Brown and Miller, 1989; Rossmanith and Miller, 1993). The studies which were reported included both analytical and experimental results.

The initial investigations of mixed mode behavior were directed toward the problem of fracture. The relationship between the mode I stress intensity factor and the energy release rate for crack extension is given by equation (2.46). Similar relations for modes II and III can be derived, and these are

$$G_{\mathrm{II}} = \frac{\left(1-v^2\right)}{E} K_{\mathrm{II}}^2,$$

(5.39)

and

$$G_{\mathrm{III}} = \frac{\left(1+v\right)}{E} K_{\mathrm{III}}^2.$$

(5.40)

If the individual energy release rates for mixed mode fracture are added, the total can then be expressed as

$$G = G_{\mathrm{I}} + G_{\mathrm{II}} + G_{\mathrm{III}}$$

(5.41)

Mode I fracture under monotonic loading is concerned primarily with the loading condition at which fracture occurs. In terms of the stress intensity factor this 'critical' value for 'elastic' fracture is $K_{\mathrm{IC}}$. For mixed mode loading a critical condition is also required, but this has been expressed as a function of the stress intensity factors for the modes involved. This is analogous to the yield condition for a multiaxial stress state or for the fatigue failure function which indicates the effect of mean stress (equation 4.3).

The crack extension plane for mode I fracture coincides with the crack plane prior to extension, i.e. there is no deviation from the original crack plane. This, of course, is based on the mathematical model for a crack and does not recognize the possibility of local, microscopic deviations. The assumption of planar extension is, however, generally representative of the dominant extension direction.

For mixed mode fracture the crack extension generally differs from the original plane, so a criterion for the direction is also required. Fig. 4.7 is an example of growth under modes I and II. Mixed mode fracture predictions, then, require the availability of two criteria, i.e. a critical crack extension equation which is a function of the individual stress intensity factors and a criterion for predicting the direction of crack extension.

Critical conditions for pure mode II and pure mode III are difficult to achieve experimentally, not only because of the possibility of friction between the sliding crack faces, but also because it is unlikely that the crack from which extension occurs is perfectly flat. Thus, both localized normal and tangential components of force can develop between the upper and lower asperities of the crack faces. As a consequence, experimental investigations of these modes generally involve mixed mode tests which are combinations of mode I and mode II or of mode I and mode III. To be valid, however, mode I must be suficient to keep the crack faces from contacting one another. It should be recognized, of course, that cracks in components cannot be subjected to this limitation, and contact between crack faces may occur.

Several rational proposals have been offered for determining the direction of crack extension from an existing crack which is subject to mixed mode loading. For one, it is assumed that crack extension occurs in a direction for which the tangential stress adjacent to the crack tip has a maximum value (Erdogan and Sih, 1963). The criterion for another proposal suggests that the crack extension direction is the one for which the strain energy density is a minimum (Sih 1974, 1991). Another suggestion is that extension occurs in a direction for which the energy release rate is a maximum (He and Hutchinson, 1989).

Palaniswamy and Knauss (1978) have indicated that the envelope for mode I and mode II fracture for the maximum energy release rate criterion can be described by the equation

$$\left(\frac{K_{\mathrm{I}}}{K_{\mathrm{IC}}}\right) + \frac{3}{2}\left(\frac{K_{\mathrm{II}}}{K_{\mathrm{IC}}}\right)^2 = 1. \tag{5.42}$$

Analogous equations can be developed for the other criteria.

For combined mode I and mode II loading, the predictions for directions and the ratios of $K_{\mathrm{II}}$ to $K_{\mathrm{IC}}$ for $K_{\mathrm{I}} = 0$ (pure mode II) are given in Table 5.2.

He and Hutchinson (1989) have also shown that a criterion based on extension being on a path along which $K_{\mathrm{II}} = 0$ predicts directions which are in close agreement with those predicted by the energy release rate criterion.

Table 5.2    Direction of crack extension

| Criterion | Direction | $K_{II} / K_{IC}$ ($K_I = 0$) |
|---|---|---|
| Max. tangential stress | 70.5° | 0.866 |
| Min strain energy | 79.2° | 1.054 |
| Max. energy release rate | 77.4° | 0.816 |

The conditions required for mixed mode crack extension tacitly assume that once a branch is started, the propagation continues to complete failure.  Pook (1983) has observed that since the extending crack path is curved, a complete solution of the problem requires that the critical conditions be re-examined continuously along the crack path. Thus, he distinguishes between criteria for branch crack initiation and criteria for continuing propagation.

When an existing planar crack is exposed to cyclic mixed mode loading, crack extension can occur by the initiation of a branch which has a direction that deviates from the plane of the original crack.  This is often referred to as a 'kinked' crack.  In an effort to provide a lower bound for the initiation of kinked cracks, some investigators have developed methods for determining what in essense is a threshold for these cyclic, mixed mode loading conditions (Tanaka, 1974; Pook, 1989; Baloch and Brown, 1993).

Pook (1989) has developed a mixed mode I and II criterion based on the observation that crack extension often occurs in a direction in which the mode II stress intensity factor at the tip of the kinked crack is zero, i.e. the kinked crack will extend in mode I.  His proposed criterion is derived from a consideration of the equations for the stress intensities, $k_I$ and $k_{II}$ at the tip of the kinked crack in terms of the stress intensities for the initial crack, $K_I$ and $K_{II}$.  These equations are

$$\Delta k_I = \cos\frac{\theta}{2}\left(\Delta K_I \cos^2\frac{\theta}{2} - \frac{3}{2}\Delta K_{II}\sin\theta\right) \tag{5.43}$$

and

$$\Delta k_{II} = \frac{1}{2}\cos\frac{\theta}{2}\left[\Delta K_I \sin\theta + \Delta K_{II}(3\cos\theta - 1)\right], \tag{5.44}$$

where $\theta$ is the angle which the kinked crack makes with the original crack. By assuming that $\Delta k_{II} = 0$, it follows that

$$\Delta K_I \sin \theta_k = -\Delta K_{II}(3 \cos \theta_k - 1), \qquad (5.45)$$

where $\theta_k$ is the kink angle for $70.5° \leq \theta_k \leq -70.5°$. The criteria for the threshold is then

$$\frac{\Delta K_I}{\Delta K_{th}} \sin \theta_k = -\frac{\Delta K_{II}}{\Delta K_{th}}(3 \cos \theta_k - 1), \qquad (5.46)$$

where $\Delta K_{th}$ is the threshold range of stress intensity for mode I.

Pook has suggested that equation (5.46) can be approximated to within one percent by the more convenient equation

$$\frac{\Delta K_{II}}{\Delta K_{th}} = \left[0.08\left(\frac{\Delta K_I}{\Delta K_{th}}\right)^2 - 0.83\frac{\Delta K_I}{\Delta K_{th}} + 0.75\right]^{\frac{1}{2}} \qquad (5.47)$$

which can be solved for $\Delta K_{II}/\Delta K_{th}$

Pook (1989) has applied this criterion to data for a number of steels and for commercially pure aluminum. He concluded that an analysis of the results indicate that equation (5.47) can be used as a conservative estimate of the threshold boundary for mixed modes I and II. The concept of the lower bound may, thus, be useful for establishing design limits.

Pook (1989) has extended the lower bound concept to include mixed mode I, II and III loading. This results in a surface with $\Delta K_{II}/\Delta K_{th}$ being the elevation above a $\Delta K_I/\Delta K_{th}$ and $\Delta K_{III}/\Delta K_{th}$ plane.

Baloch and Brown (1993) have conducted mixed mode I and II experiments on A1S1 316 austenitic stainless steel and BS 4360 50D structural steel and have found that the data obtained provides support for the lower bound concept. They observed, however, that closure obstruction in the form of surface roughness indicates that effective values of $\Delta K_{II}$ should be introduced. Also, in contrast to mode I experiments, they observed that for their mixed mode tests, the crack surfaces remained in contact with one another even for high $R$ values. This closure obstruction feature is examined again later in this section.

An understanding of the complexity of the mixed mode problem can be appreciated by an examination of two of the assumptions which are made by Pook (1989) in his development of the lower bound concept. These are: the ratios of the stress intensity factors and the stress ratio, $R$, do not vary during a fatigue cycle; and the planes of the initial crack and the kinked crack are flat and the crack fronts are straight. Although these conditions may be approximated for some combinations of mixed mode loading, they are questionable for the three dimensional mixed mode loading of surface cracks. Here, the ratios can vary along the crack front and the crack extension surface can be curved.

An example of the development of a three dimensional crack surface is present in an article by Tohgo, Otsuka and Yoshida (1990), who present results which are of general interest in the development of criteria for crack extension orientation. They conducted two distinct fatigue crack experiments in which the ratios of $K_{II}$ and $K_{III}$ varied differently along elliptical crack fronts. The tests were designed to exclude fracture surface contact, i.e. the cracks were open during cycling, so closure was not involved. In each of the experiments, the orientations of the cracks relative to the loading differed for the two types of experiments. Tests were conducted on two alloys: SM41A structural steel and 2017-T4 aluminum alloy.

Two mixed mode fatigue crack experiments were conducted on each metal. The mixed mode loading began after surface cracks were introduced, and in case 1 mode II was dominant at the middle of the crack front and in case 2 mode III was dominant at the middle of the crack front. The latter case is similar to that depicted in Fig. 2.8. Photographs of the fatigue crack surfaces are shown in Fig. 5.19 (a), (b), (c) and (d). Case I loading results are shown in Fig. 5.19 (a) and (b), and case 2 results are shown in Fig. 5.19 (c) and (d). In each case the initial, planar pre-crack is the inner semi-circular area, and the outer, annular surface is the growth after the introduction of mixed mode loading. As can be seen, the annular surfaces for the steel and the aluminum differ for each loading case. Only in Fig. 5.19 (b) and (d) do the extended crack surfaces coincide with the pre-crack plane. The extended surfaces in Fig. 5.19 (a) and (c) are three dimensional.

(a)

(b)

(c)

(d)

**Fig. 5.19** Fatigue crack growth under mixed mode II and III loading (Tohgo, Otsuka and Yoshida) (1990)

For the steel, cracks were observed to grow on planes upon which a maximum tensile stress was acting. For the aluminum, growth was on planes upon which the shear stress was a maximum. This indicates that criteria for the orientation of fatigue crack extension may have to consider not only the stress state at the crack tip, but also the character of the preferred mode of response of the given material.

Efforts to develop methods for predicting the path of an extending crack under mixed mode I and II loading have focused on the use of Paris type equations for cyclic growth rate. Linnig (1993) has presented experimental and analytical results for the aluminum alloy 7075-T651 and the plastic PMMA. His growth rate equation is of the form

$$\frac{da}{dN} = f(\Delta K_{eff}),$$
(5.48)

where $\Delta K_{eff}$ is of a form similar to one derived by Sih (1974). A finite element program was used to determine the angle of the initial kinked crack and the curved path of the extending crack. He obtained agreement between calculated and experimental results for both the initial kink angle and the curved path except for cases in which the ratio of $K_{II}$ to $K_I$ exceeded about 0.7. Above this value the experimental kink angles were less than the calculated angles. He attributed the difference to closure obstruction in the form of friction and surface roughness.

Badaliance (1981) has reported the results of mixed mode I and II loading on a 7075-T735 aluminum alloy specimen with a crack inclined to the loading direction. He used a cyclic growth rate equation of the form

$$\frac{da}{dN} = f(\Delta S),$$
(5.49)

where $\Delta S$ is the strain energy density factor range proposed by Sih (1974). For a specimen with an initial crack inclination angle of 45°, coordinate values along the experimental and calculated paths were within 10% of one another after 2950 and 4800 cycles of loading.

For the types of equations used for the examples cited, the effects of obstruction to closure are not accounted for. Because of this limitation, the use of the lower bound concept proposed by Pook (1989) appears to provide a conservative, but less uncertain basis for predicting mixed mode crack growth behavior.

For the examples of mixed mode loading which have been described, the local stress state at a crack tip depends only on the specimen geometry and the loading details. When obstruction to closure is present, the local crack face loading can contribute to the crack tip stress state. It follows that the stress intensity factor for such cases has both a global component and a local component. Forsyth (1983) has presented results which emphasize the importance of microstructural features in the formation of fracture surfaces. He suggested that crack front advance considerations should address the development of microscopic as well as macroscopic features. The use of global and local components of the stress intensity factor is compatible with this observation.

Micrographs of fracture surfaces, such as those shown in Fig. 5.2, indicate that there are jogs or steps along the crack path. Misfits between inclined facets can, therefore, introduce local mode II loading conditions during closure even though the global loading is of the mode I type. A model which possesses both global and local loading features has been developed (Carlson and Beevers, 1985), and the results indicate that the inclination angle of a facet step and the friction coefficient affect the values of the stress intensity factors. While the crack is open, only a global $K_I$ is developed. Both $K_I$ and $K_{II}$ are present, however, during closure contact. Because of the presence of friction, moreover, the loading and unloading paths differ. This introduces a local energy absorption hysteresis loop during the closure part of the loading cycle. It is also of interest to note that during unloading, $K_I$ decreases and $K_{II}$ increases. Thus, since the ratio of the stress intensity factors is not constant, non-proportional loading conditions are developed adjacent to the crack tip. In reference to the previous discussion regarding compressive excursions, the increase in $K_{II}$ with decreasing external load could be expected to continue when the loading becomes compressive. This increase in the local $K_{II}$ may contribute to observed effects of compressive excursions. The results presented suggest that though the global stress intensity factor may dictate the orientation of the macro-fracture plane, conditions for local branching on a micro-scale may be dictated by the combined global and local state.

Another combination of global and local loading states has also been observed in circumferentially notched cylindrical bars subject to torsion. Here circumferential cracks have been observed to develop in tests designed to investigate mode III fatigue crack growth. Because the fracture surface is not flat, the surfaces 'ride up' relative to one another and produce a local mode I condition (Tschegg, 1983, Tschegg, et al. 1992). It is also clear that friction would be a factor in this example.

A considerable amount of effort has been expended on the subject of fatigue crack growth under mixed mode loading conditions. Although progress in identifying the basic features of the problem has been made, a number of issues remain to be resolved. Predictive codes designed to incorporate the effects of mixed mode loading can, therefore, be expected to possess a greater degree of uncertainty than those associated with mode I growth. Mixed mode experiments on actual components, under service loading conditions, should ultimately be used to evaluate predictive methods which have been proposed.

## 5.7 ENVIRONMENT

### 5.7.1 Introduction

Extremes of temperature – low and high – and active atmospheres can result in a change in the character of the operative mechanisms which affect fatigue crack growth. Exposure to low temperatures occurs in ocean-going ships and other vehicles in arctic environments and in storage facilities for liquefied gases. High temperatures are encountered in gas turbine engines and stationary power generation equipment.

The effect of the formation of oxides on the development of closure obstruction is referenced in section 5.2.2. Other effects of corrosive environments which lead to pit formation, cracking and crack growth are quite complex. The interaction of corrosion and fatigue can be devastating. A consideration of the consequences of this type of damage in aircraft which have had extended service lives has generated substantial activity in this area of investigation.

## 5.7.2 Low temperatures

The sensitivity of the strength properties of metals to low temperature exposure depends on crystal structure. The strength and ductility properties of body-centered cubic metals, such as ferritic alloys, are quite sensitive to exposure to low temperatures. This is manifested in a behavior in which a reduction in temperature below a 'transition temperature' results in a change from ductile to brittle fracture. The fracture characteristics of face-centered cubic metals, such as aluminum, nickel, copper and austenitic steel alloys, are much less susceptible to drastic changes at low temperature.

Tschegg and Stanzl (1981) have conducted fatigue crack tests on both BCC and FCC metals at 77 °K and 293 °K. For the FCC metals – copper and 304 stainless steel – they found that the rate of crack growth for a given $\Delta K$ was smaller for the lower temperature. The threshold values of the $\Delta K$ were about 10% and 20% smaller for the higher temperature for copper and the 304 steel, respectively. The fracture surfaces were ductile, transcrystalline for both metals for all test conditions. All tests were conducted under $R = -1$ loading and $\Delta K = K_{max}$ was used in the correlations of data, i.e. the effect of the compressive part of each cycle was ignored (section 5.3.5). Since the yield strength of these metals was greater at the low test temperature, the amount of crushing of closure obstructions during compressive loading would be expected to be greater at the higher test temperature. If this effect were accounted for, there could be a decrease in the difference between the rate curves for the two temperatures. It is of interest to note also that differences in yield strengths for the two test temperatures is greater for the 304 steel, which appeared to have the greater growth rate differences between the high and low test temperatures. Thus, the actual differences, due to temperature, may be negligible for these metals.

For the BCC metal – a low carbon steel – the rates again were observed to be greater for the higher temperature for a given value of $\Delta K$. Also, the threshold value at 293 °K was about 40% lower than that for tests at 77 °K. At 293 °K, the fracture surfaces were ductile, transcrystalline. At 77 °K, the fracture was of a brittle, cleavage character for growth rates greater than $10^{-8}$ mm per cycle and ductile transcrystalline for rates less than $10^{-9}$ mm per cycle. This indicates that the fracture surface developed is dependent on both the growth rate and the stress level.

The fatigue crack growth results reported by Tschegg and Stanzl (1981) are generally consistent with the differences in the responses to low temperatures of BCC and FCC metals.

### 5.7.3 Elevated temperatures

When the ratio of the absolute value of the exposure temperature to the absolute melting temperature (the homologous temperature) of a metal approaches or is greater than 0.5, significant time-dependent, inelastic flow or creep can occur. Cyclic loading at an elevated tempeature can, therefore, result in both creep and fatigue damage. The mechanisms of fatigue and creep differ, and interaction or coupling effects can be expected to be operative.

A number of variables affect the type of interaction which occurs. At very high tempeatures, creep crack growth can become dominant. At the low range of temperatures, creep crack growth can be of secondary importance. Cyclic frequency can be expected to be a factor, particularly at the high range of temperatures. In the operation of gas turbine engines a maximum load 'dwell' time is encountered and it must be given consideration. Finally, the effect of the atmosphere can also be a factor. An active atmosphere can introduce various forms of corrosion which can interact with creep and fatigue mechanisms.

The mechanisms which are operative in fatigue crack initiation and growth are discussed in section 3.2. These mechanisms have been found to be growth rate dependent, and they differ in three distinct regions: the near threshold region the intermediate Paris region and the accelerating growth region.

Mechanisms which are operative in creep involve diffusion processes which are temperature and stress dependent. These include dislocation movement and the migration of vacancies and atoms. These mechanisms have also been used to explain grain boundary sliding which is a high temperature deformation mechanism (Hertzberg 1989). Also, Raj and Baik (1980) have discussed a mechanism in which cavities are formed in front of a crack tip and crack growth occurs by the successive rupture of ligaments between cavities.

Deformation maps which are plots of stress versus the homologous temperature have been used by Ashby (1972) to partition the plots into separate zones within which various high temperature mechanisms are operative. Since grain boundaries play a role in the deformation mechanisms, the maps for two different grain sizes of the same material can differ.

Although both fatigue and creep are stress dependent, creep, unlike fatigue, is also temperature and time dependent. The operative mechanisms are, therefore, different. In creep crack growth a 'process' volume in front of the crack tip is exposed to high stresses which, by creep deformation mechanisms, locally alter the condition of the material. The changes which occur, however, differ from those in the cyclic plastic zone in fatigue. The creep process zone and the fatigue cyclic plastic zone are superimposed upon one another during creep – fatigue crack growth.

At the lower part of the temperature range for creep, and for small stresses and high cyclic frequencies, crack growth is cycle dependent and can be characterized by $\Delta K$. At the other extreme of high temperatures, high stresses, and low cyclic frequencies, crack growth is time dependent. Several parameters have been proposed to characterize the state adjacent to the tip of a crack in a body subject to creep. These are based on the use of an elastic nonlinear viscous constitutive law which expresses strain rate components in terms of the stress components.

One of the proposed parameters, $C(t)$, is based on the use of a nonlinear viscous law in a contour integral evaluation which is analogous to the $J$ integral evaluation (Bassini and McClintock, 1981). If a steady state condition in which the stress state becomes fixed in time is attained, $C(t)$ becomes $C^*$. Landes and Begley (1976) have expressed $C^*$ as a power dissipation parameter which can be evaluated experimentally from measurements on creep crack growth specimens.

Gieseke and Saxena (1989) have proposed the use of a parameter, $C_t$, which can be determined by measuring the power release rate during creep crack extension. Yoon, Huh and Saxena (1993) have also evaluated $C_t$ for an elastic – plastic nonlinear viscous solid.

The intermediate condition between the extremes of cycle and creep dominance is of considerable engineering importance. One of the proposals offered to analyze this case is to use a simple superposition of fatigue and creep crack growth (Saxena, 1988). This results in the relation

$$\frac{da}{dN} = \left(\frac{da}{dN}\right)_{\text{fat}} + \left(\frac{da}{dN}\right)_{\text{cr}} \tag{5.50}$$

The use of equation (5.50) assumes that linear superposition is valid and that there is no coupling or interaction between creep and fatigue mechanisms. Also, the computation of the two growth rate components is based on the use of constitutive laws which are determined by tests on standard laboratory specimens. A question regarding coupling effects arises on the use of such test data to describe the properties of the small volume of material in front of the crack tip. Do, for example, the deformation mechanisms of fatigue alter the creep properties, and vice versa. The extent to which these questions pose problems must be resolved by an evaluation of applications of analytical methods to experimental results.

### 5.7.4  Corrosion – fatigue

A comparison of the fatigue resistance of metals and alloys in aggressive and inert environments usually reveals that the active environments have a detrimental effect. Active environments include either gaseous or aqueous atmospheres. Of these atmospheres the aqueous is generally found to be the more aggressive. This excludes such exposures as the very high temperature, catastrophic oxidation of such materials as graphite, tungsten and columbium in air. These are avoided in engineering applications.

Simultaneous exposure of a metallic surface to an aggressive aqueous solution and constant loading can result in an embrittlement which leads to surface cracking. This phenomenon is described as 'stress – corrosion cracking'. Cyclic loading can accelerate both the initiation and the growth of cracks and the progress of this type of damage is described as 'corrosion fatigue'.

In aqueous environments, sites from which cracks can nucleate have been found to be small pits. These form by selective corrosion at slip steps, grain boundaries, inclusions and locations where a protective oxide film has cracked. The fundamentals of pitting corrosion are discussed in a text by Heitz, Henkhaus and Rahmel (1992).

In a gaseous atmosphere such as air, it has been suggested that the formation of oxides can inhibit slip reversibility under cyclic loading and thereby introduce jogs in the surface. These surface irregularities can serve as crack nucleation sites. Hydrogenous gases can lead to hydrogen embrittlement and brittle cracking.

The mechanisms associated with crack nucleation and growth by corrosion differ from those for cyclic loading. The combination of cyclic loading and corrosive attack, therefore, results in the superposition of two distinct sets of mechanisms. The damage which accumulates in front of the growing crack can involve interaction between these mechanisms. The consequences of this combination of distinctive sets of mechanisms, acting in concert, results in what is sometimes described as a 'synergistic' behavior.

Under a fixed applied load, a crack can grow in a body which is exposed to an aggressive environment if the level of loading is sufficient. By use of linear fracture mechanics, the loading state can be expressed in terms of the stress intensity factor. The value of the factor which must be exceeded for growth to occur has been designated as the static stress intensity factor, $K_{ISCC}$. This corresponds to the value of a subcritical crack (SCC) below which a crack will not propagate. Above this value, crack growth rate increases rapidly during an initial phase, reaches a constant growth rate in a second phase, and increases rapidly in a third phase as the value of $K$ approaches $K_{IC}$. The general features of this behavior are depicted in Fig. 5.20.

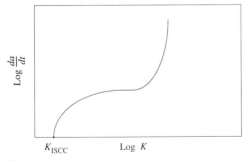

**Fig. 5.20** Crack growth rate dependence on stress intensity factor in a corrosive environment

McEvily and Wei (1972) have suggested that the relationship between the cyclic crack growth rate and the stress intensity factor depends upon the value of $K_{ISCC}$ relative to the value of $K_{max}$. They suggest that if $K_{max}$ is less than $K_{ISCC}$, the rate of growth is increased in the near threshold and Paris regions as the crack grows through the embrittled crack tip volume. Note that no growth would have occurred under a constant load for this condition. An abrupt growth rate increase can, however, occur when $K_{max}$ becomes equal to and greater than $K_{ISCC}$. (Remember that for $K_{max}$ greater than $K_{ISCC}$, growth could occur even under a constant $K$). This suggests that the combined effects of these mechanisms may result in substantial rate increases in both the near threshold and Paris regions. The end of third region, in which $K_{max}$ approaches $K_{IC}$, is unchanged in this interpretation. Corrosion is time-dependent and in this final stage, the crack growth rate is increasing rapidly. The explanation suggested by McEvily and Wei (1972) is depicted in Fig. 5.21.

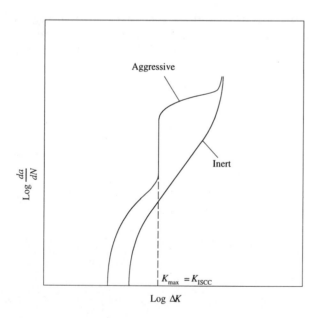

**Fig. 5.21** An interpretation of fatigue crack growth in an aggressive environment (McEvily and Wei, 1972)

Bucci (1970) and Wei and Landes (1970) have proposed the use of a superposition model to account for the combined effects of cyclic loading and corrosion. They suggested that the rate of crack growth could be expressed as a linear summation of the form

$$\frac{da}{dt} = \left[\frac{da}{dt}\right]_f + \left[\frac{da}{dt}\right]_{sc}. \tag{5.51}$$

The first term represents the fatigue crack growth rate which would have occurred in an inert environment. The second term represents growth due to corrosion, and it must incorporate the effects of cycle frequency and loading wave forms. The fatigue contribution can be related to the cyclic loading history through use of the relation

$$\frac{da}{dt} = \frac{da}{dN}\frac{dN}{dt}, \tag{5.52}$$

where the actual loading history must be used to determine the cyclic growth rate for exposure in an inert environment.

The components in equation (5.51) are obtained from tests which do not include coupling effects. These may modify the growth rate contributions of the two activities, e.g. the crack is growing into a small volume of material which has absorbed both cyclic loading and corrosion damage.

Wei and Simmons (1981) have proposed a modification of equation (5.51) by including an additional term which is intended to provide a contribution from the joint effects of cyclic loading and corrosion. The modified equation is still, however, based on linear superposition.

McClintock (1972) has presented results of a fundamental investigation of corrosion fatigue and has listed variables upon which crack growth rate is dependent. Among the fracture mechanics variables cited are the geometrical features in the neighborhood of the crack tip. These are the size and shape of the plastic zone, the crack tip opening displacement and the crack tip opening angle. The latter two can be expected to have an influence on the access of a corrosive medium to the crack tip. At small crack growth rates, the mechanics of microscopic crack branching may also be a factor.

For corrosion fatigue McClintock has identified variables which are associated with the corrosion process. These include the cyclic frequency and quantities which are dependent variables in equations for diffusion, electrical conductivity and film growth. The corrosion fatigue problem thus formulated consists of a system of coupled differential equations.

Though the theoretical description of the problem is comprehensive, it still does not account for the corrosion assisted mechanism associated with the progressive, microscopic loss of coherency in front of irregular crack fronts. After discussing the complexities of the coupled corrosion fatigue problem, McClintock (1972) concludes that a thorough analysis of well designed experimental investigations would be more expedient than purely analytical studies.

The complexity of the problem of fatigue crack growth in a corrosive environment can be further illustrated by reference to experimental results reported by McClintock (1972). Crack growth experiments were conducted on aluminum alloy 7075-T6 specimens in a 3% solution of NaCl at 1 Hz. To examine the effect of cathodic protection (Heitz et al., 1992), tests with applied electrical potentials of – 0.70 and –1.20V were performed. A plot of the data for the rate of crack growth versus the range of stress intensity factor is presented in Fig. 5.22. Note, incidentally, that by plotting the data on an undistorted, cartesian coordinate system, an interpolation between zero growth and the first data points can be performed. Contrary to expectation, the growth rates for the specimen with the greater cathodic protection (– 1.20V) are larger than those for the specimen with lesser protection (– 0.70V).

Striation spacings on the fracture surfaces were measured for the two cathodic protections, and it was found that the values of the spacings were of the same order of magnitude. The surface roughness was measured in terms of 'wave lengths'. The average wave lengths were $25 \mu m$ and $40 \mu m$ for the – 0.7V and –1.20V tests, respectively. The surface with the lesser cathodic protection (– 0.7V) was observed to be smoother but exhibited more branch cracking along the crack front. Branch cracking can be expected to increase resistance to crack growth.

McClintock cites other examples in which corrosive exposure can result in an irregular crack front which increases the resistance to crack growth. Chu and Wacker (1969), for example, found that the residual fracture toughness of aluminum alloy 7079-T6 specimens exposed to sea water was greater than that of unexposed specimens. Fractographic examinations revealed that the crack fronts of exposed specimens were more irregular than those of unexposed specimens.

Consideration of the results presented by McClintock (1972) clearly indicates the complexities of corrosion fatigue behavior, and reveals the importance of conducting experimental investigations.

A relatively simple experimental method has been developed for assessing the resistance of metals to crack growth in corrosive environments (Bucci et *al.*, 1986). These investigators have proposed a method in which the residual strengths of exposed specimens are determined. Reductions in strength for different periods of exposure are found by comparing residual strengths with the strengths of unexposed specimens. The procedure provides an accelerated method of discriminating degrees of susceptibility to degradation, and the fracture data can be characterized by use of extreme value statistics (Weibull 1939, Epstein 1948). Fractographic examination of specimens exposed for different times provide estimates of the effective flaw sizes which are developed (section 6.2.2). By representing the flaws by equivalent cracks with geometries which can be analyzed by fracture mechanics methods, correlations based on the theory of linear fracture mechanics and on elastic – plastic fracture mechanics were shown to be possible.

An application of the method to an assessment of corrosion resistance for three temper conditions of the aluminum alloy 7075 indicated that stress corrosion cracking ratings were consistent with accepted but more time consuming methods. Although the method as described entails the use of a constant applied stress during exposure, the authors observe that it could be adapted for evaluations involving corrosion fatigue and wear.

It should be recognized that if corrosion interactions are incorporated in a fatigue crack growth predictive code, the effects of closure obstruction should also be included. Obstruction to closure reduces the effective range of the stress intensity factor and can, therefore, provide a beneficial counter-effect to the detrimental action of corrosion. It is, for example, important to note that oxide film formation, though a corrosion mechanism, can, by obstructing closure, be beneficial. Finally, it should also be realized that cracks initiated by corrosion fatigue are initially 'small'. As a consequence, the issues arising from the small crack growth behavior discussed in section 5.4 must be added to those which have been considered in this section.

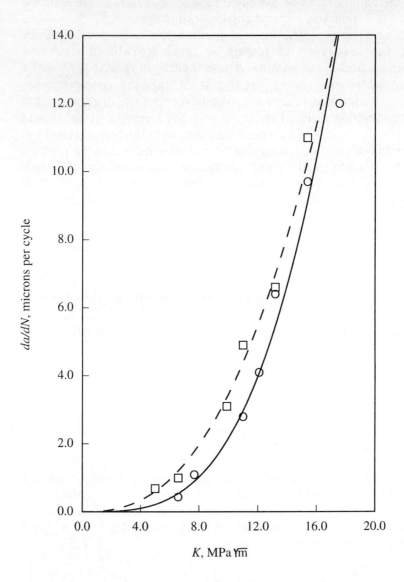

**Fig. 5.22** Crack growth rates for aluminum alloy 7075-T6 in a 3% solution of NaCl at 1 Hz for – 0.70 V ○ and for – 1.20 V □ (McClintock, 1972)

## 5.8 NONMETALS

### 5.8.1 Brittle solids

Data on crack growth has been obtained for brittle materials and plots of crack growth rate versus range of stress intensity factor have been reported. As noted in section 3.4, however, the Paris equation exponents have generally been greater than ten. This indicates that very small changes in stress intensity factor range can result in large changes in crack growth rate. These materials are not, therefore, damage tolerant. It should also be noted that the growth which has been observed could be described as a localized, progressive comminution, rather than a crack growth.

Ultimately, the choice on the type of design approach which is appropriate depends upon the nature of the application which is being considered. This has been nicely put into perspective by McClintock (1972), who has observed that 'If a failure is likely to cause fatalities or extreme expense, it is probably advisable to postulate initial flaws and choose stress levels and inspection procedures to assure that cracks will be removed from service before they become critical. In other applications where failure is not catastrophic, production is reliable, and the cost of downtime and replacement not too high, I should think it would be quite satisfactory to design on initiation (e.g. an unnotched S – N curve) and accept occasional failures'.

In view of the damage growth behavior of brittle materials it may be concluded that design applications based on safe life rather than crack growth would be preferable. Also, because of uncertainties cited in section 4.5.2, actual testing of proposed components may be advisable. Crack growth tests are of value in the identification of damage mechanisms. Also, they can be useful in evaluating improvements due to fabrication changes and in providing comparison data for materials being considered for specific applications.

### 5.8.2 Polymeric solids

Crack growth in metals at temperatures well below one half of the homologous temperature is primarily dependent on the stress intensity factor, with adjustments to account for threshold, mean stress, critical stress intensity factor and closure obstruction. Crack growth laws such as equation (5.4) can, therefore, be used as a basis for predicting crack growth histories. In addition to the above variables, crack growth in polymeric materials can also be dependent on cyclic frequency. As a result, the localized region in front of a crack tip can be subjected to an accumulation of both thermal and mechanical damage (section 3.3).

When there is a strong dependency on frequency, crack growth predictions can, therefore, be expected to be difficult. The results of Hertzberg, Manson and Skibo (1975) on polystyrene and cross-linked polystyrene (section 3.3) indicate that frequency effects can be ameliorated by a judicious choice of materials.

A large number of applications which make use of polymeric materials do not involve exposures to high frequency loading. In such cases design based on static strength is appropriate. Where cyclic loading is a factor, and a dependency on frequency is not involved, one option is to limit the stress range to values below the threshold for crack growth. An alternative is to use a safe life approach, and to base design stress values on stress versus cycles to failure diagrams.

## 5.9 COMPOSITES

Among the three composite matrices which have been considered, the polymeric matrix has received the most attention and has been the most widely used. For this reason the fatigue crack growth behavior of composites with polymeric matrices has been made the sole topic of Chapter 7. Ceramic and metal matrix composites are considered in this section.

The creation of ceramic matrix composites is largely motivated by a desire to provide a class of materials which can be used for very high temperature service. Ultimate failures can, therefore, be a consequence of a coupling of fatigue and creep damage mechanisms. The use of Paris type crack growth rate laws are, hence, questionable as a basis for component design. Also, from the relatively large values of the Paris exponents which have been reported (section 3.7.2), it may be concluded that the crack growth behavior of the reinforced ceramics is similar to that of the unreinforced ceramics, i.e. very small changes in load can result in large changes in crack growth rate.

The statistical character of the initial flaw distribution and the subsequent damage in the form of microcracking during loading introduce uncertainties which are difficult to assess. Thus, although crack growth tests can be of value in evaluating the effectiveness of crack tip shielding mechanisms, they cannot as yet be used for predicting crack growth in components made for service applications. Actual component testing, therefore, provides a sounder basis for the evolution of designs than analyses of the results of standardized crack growth tests.

Metal matrix composites are fabricated in a wide variety of forms (section 3.5.1). The type of damage that is encountered depends both on the details of fabrication and on the loading conditions. One common property, however, has a strong influence on the characteristics of induced damage. This property is the interfacial bond strength, and it plays a dominant role in the behavior of metal matrix composites. The damage mechanisms developed in adhesively bonded laminates and fiber-metal laminates, as well as composites with filament, whisker and particulate reinforcements are all dependent on interfacial bond strength. Delamination and debonding processes can, by the introduction of crack branching and bridging mechanisms, absorb stored strain energy and increase resistance to crack extension.

Although much mechanical property data have been obtained on the many types of metal matrix composites which have been proposed, the objectives of the studies have generally not been to acquire data for the design of structural components. In many investigations the objective has been to determine how fabrication variables affect the stiffness, strength and fatigue properties.

To be useful in a service application a candidate material must have been developed to a degree which makes it possible to manufacture the materials in useful quantities. Further, because of the complexities of damage mechanisms and the need to integrate components into a structural system, component testing is advisable at the present state of developments. Basic data are necessary for evaluation and screening of these composites, but ultimately their use in components should be evaluated in tests which simulate the proposed operating conditions.

An appreciation for the need for the recommended procedure is nicely revealed by reference to two component type tests described by Schijve (1993). In one, skin panels with four stiffeners were produced entirely by the use of a fiber – metal laminate. Cutting, forming, riveting, and bonding, which required a second curing cycle, were required to fabricate the panel. Fatigue testing revealed the location, extent and character of the damage incurred. Both crack growth and some delamination were discovered, but the panel was considered to be structurally acceptable after completion of the tests.

For a second test a fiber – metal composite lug, which was identical in form to one used in an aircraft in service, was fabricated. Simulated flight loading equivalent to 92000 flights at the maximum load level was applied without failure. It was concluded that the test demonstrated that the part had a large safety margin and would be operationally sound for use with long inspection intervals.

The tested materials in these examples are commercially available, and they had in the course of development undergone intensive evaluation. The investigations were, however, considered to be necessary for demonstrating the adaptability of these materials for actual components.

# 6

# Structural integrity of metals

## 6.1 INTRODUCTION

The use of the concepts of damage tolerance has been briefly referred to in previous sections. The elements are now dominant both in the design of new aircraft and in the re-evaluation of 'aging' aircraft for which the original construction was based on the safe life design strategy.

Often during the development of a new technology a specialized terminology which facilitates efficient communication evolves. A comprehension of terms which are frequently used in the relevant literature is, therefore, essential in an introduction to the subject. In the discussion which follows the primary emphasis is on fatigue crack growth in aircraft airframes. This includes the fuselage, empennage, landing gear, control systems, engine mounts, etc.

'Damage tolerance' refers to the capability of an airframe to resist failure originating from flaws or cracks for a predetermined service life.

'Durability' is the ability of an airframe to withstand cracking, corrosion, thermal degradation and wear for a prescribed period of service life.

When the extent of damage and repair costs has reached an unacceptable level, the 'economic life' has been exceeded.

A 'structural integrity program' has the objective of establishing, evaluating and confirming the strength, rigidity, damage tolerance and durability of a given aircraft.

The subject matter discussed in this chapter evolved during a period in which fracture mechanics was extended to describe fatigue crack growth in metals. The use of the analytical methods described requires the utilization of crack growth laws which can be applied to predict stable crack growth. Computer codes have been developed for the necessary computations and tabulations of the required data on commercially available alloys are obtainable. Although some issues regarding behavioral responses have not been completely resolved, the basic foundation for the application of crack growth methods to the damage tolerance analysis of metallic components is well established.

Although the characterization of fatigue damage in nonmetals and composites is well advanced, a general consensus on the applicability of damage tolerance concepts to these materials has not been achieved. The topics presented in this chapter begin with a discussion of analyses involving metallic materials. A discussion of some of the emerging proposals for the use of damage tolerance concepts in polymeric matrix composites is presented in Chapter 7. The structural integrity of surgical implants (biomaterials) is discussed in Chapter 8.

Since monitoring stable fatigue crack growth is a crucial part of the damage tolerance philosophy, inspection schedules for the nondestructive examination of selected sites must be established. A variety of techniques are available for examining metals, nonmetals and composites for flaws, cracks and delaminations. A comprehensive treatment of nondestructive testing techniques is presented in volumes I and II of the *Nondestructive Testing Handbook* which was edited by McMaster (1959). A good introduction to the principles of non-destructive testing is presented in a text by McGonnagle (1961). Summaries of currently used techniques are presented in a handbook edited by Bøving (1989). An informative listing of techniques which can be used and of the types of flaws which can be detected are contained in tables on pages 4 and 5 of this reference. The probability of the detection of flaws of a given size depends upon the accessibility of the inspected site and the circumstances of the inspection, e.g. in a depot, in the field, during fabrication. The type of technique selected and the sensitivity of detection can be influenced by these factors.

## 6.2 STATISTICAL CONSIDERATIONS

From the preceding definitions it is clear that methods for predicting the initiation of cracks from pre-existing flaws and the subsequent crack growth are required for the attainment of structural integrity. Although the methods of crack growth analysis which have been described in previous sections have implicitly been deterministic in character, it is clear that methods for predicting durability in airframes and fleets of aircraft should have a statistical basis. The complexities of the problems which are involved introduce numerous elements of uncertainty. For a given type of component a critical location for crack initiation can often be determined, but the severity of the initial flaw from which the crack is initiated can vary from component to component. There will then be a distribution of pre-existing flaws and it follows that the crack initiation histories will differ.

### 6.2.1 Estimation of risk

Cracks, once nucleated, will initially be small, so the probability of detection will be small. Only after the cracks have grown sufficiently will a high level of the probability of detection be attained. Lincoln (1985) has described a risk assessment procedure which makes use of probability distributions of crack size and stress state along with a probability of crack detection function to determine the single flight probability of failure. The crack distributions used to illustrate the procedure were based on inspections of critical locations in retired trainer aircraft wings. The probabilities of exceeding a given stress at each critical location were derived from the stress exceedance function for the aircraft. Additional input included the critical crack length versus stress and the crack length histories for each location.

The joint probability density function is the product of the crack length and stress probability density functions. The single flight probability of failure is then the volume which is under the joint probability surface and outside of the boundary for the critical crack length versus stress curve. It should be noted that as crack length increases with increasing flight hours, the joint probability function also changes, so the computed risk reflects the changes which occur during the service exposure.

Lincoln (1985) also observed that the results described could be extended to include a determination of the probability of failure of a group of aircraft or the expected number of losses from a fleet of aircraft. The procedure has, therefore, potential use as a risk management tool for a fleet of aircraft. He also describes how the probability of failure is modified by periodic inspections and suggests that risk assessment can be used to make decisions for inspection intervals and the need for design modifications. Procedures for performing risk analyses have been the subject of continuing investigation and Berens, Hovey and Skinn (1991) have developed comprehensive methods for the risk management of aging aircraft.

### 6.2.2 Equivalent initial flaw sizes

A deterministic procedure for determining crack growth ignores the fact that there can be uncertainties in both the crack initiation and crack growth phases of the process. The associated uncertainties may be included in a crack growth model in several ways. Possible combinations which describe the phases as stochastic processes are described in Table 6.1, where $a(0) = a_0$ represents the initial, equivalent flaw size and $a(n) = a_n$ represents the crack length after $n$ cycles or flights.

Table 6.1  Crack growth models

| $a(o)$ | $a(n)$ |
| --- | --- |
| deterministic | deterministic |
| stochastic | deterministic |
| stochastic | stochastic |

The first combination of these models indicates that both $a(o)$ and $a(n)$ have only single values. For a selected value of $a(o)$ the value of $a(n)$ could, for example, be determined by use of the Paris law. The growth history for this method of analysis is shown in Fig. 6.1.

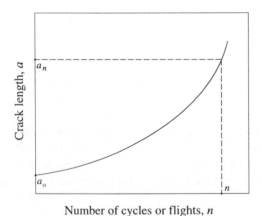

Fig. 6.1  Deterministic crack growth model

The most general combination is the fully stochastic model which is depicted in Fig. 6.2. Here, the initial, equivalent flaw sizes are represented by a distribution density function. Each value of the initial distribution function can, moreover, result in a distribution of crack paths or histories. After a period of cyclic loading, there will, therefore, be a density distribution function which represents the statistical spread in $a(t)$ values. Note that each possible initial value can, after $n$ cycles, be represented by a distribution function.

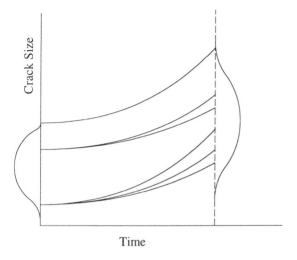

**Fig. 6.2** Fully stochastic crack growth model

The stochastic modeling of crack growth has been extensively used to develop methods for the analysis of durability (Manning and Yang, 1987). These investigators have described a variety of models in their articles. Only elementary examples of these are described here, but the interested reader may refer to the listed references for comprehensive expositions.

Fatigue is usually described as a process involving two phases: crack nucleation and crack growth. From the discussion of small cracks in section 5.4, it may, however, be concluded that a third phase may be identified. The behavior which characterizes 'small' crack growth is distinct from that of crack nucleation and the growth of 'long' cracks. The objective of the analytical methods developed by Manning and Yang (1987), however, is to construct a crack growth model which encompasses all of these phases. These investigators have focused on the initial fatigue quality of, for example, a fastener hole, and represented that property by an equivalent initial flaw size distribution function (EIFSD), $p_{a(0)}(x)$. This essentially proposes that a stochastic representation of the effective initial conditions can be used as a basis for describing the complete evolution of the fatigue process.

The features of the equivalent initial flaw size have been characterized by the following specifications:

1. It is an artificial quantity which represents the initial quality at the potential fatigue failure site.
2. The concepts of fracture mechanics are applied, but no distinction is made between small and long cracks.
3. Available fractographic crack size data along with a crack growth law are used to extrapolate backwards to the initial state.

A backward extrapolation procedure is used to determine the parameters of an equivalent initial flaw size distribution function. For a two parameter, Weibull compatible model the cumulative distribution function used has the form

$$P_{a(0)}(x) = \exp\left\{-\left[\frac{\ln(x_u/x)}{\phi}\right]^\alpha\right\} \tag{6.1}$$

for $0 < x \le x_u$ and $P_{a(0)}(x) = 1$ for $x \ge x_u$. Here, $x$ is the crack size, $x_u$ is the upper bound value of $x$, $a(0)$ is the effective crack size at time zero, and $\alpha$ and $\phi$ are Weibull parameters. The crack growth rate equation used has the form

$$\frac{da}{dt} = Q[a(t)]^b \tag{6.2}$$

where $Q$ and $b$ are empirical constants. The crack length $a(t)$ can be determined by integration of this rate equation.

The cumulative crack size distribution function, $P_{a(t)}(x)$, for a time $t$ can be obtained by integrating equation (6.2) and substituting the resulting $a(t)$ into equation (6.1). For a value of $b = 1$ the cumulative distribution function has the form

$$P_{a(t)}(x) = \exp\left\{-\left[\frac{\ln(x_u/x) + Qt}{\phi}\right]\right\}$$ (6.3)

for $x > 0$ and $t \geq 0$. The evolution of the crack size distributions is illustrated in Fig. 6.3, where for this example $x_1 = x \exp(Qt)$. The shaded areas represent the cumulative distributions of crack size for times $t = 0$ and $t > 0$. The probability that a crack size will exceed the value $x_1$ at $t > 0$ is given by $[1 - P_{a(t)}(x_1)]$.

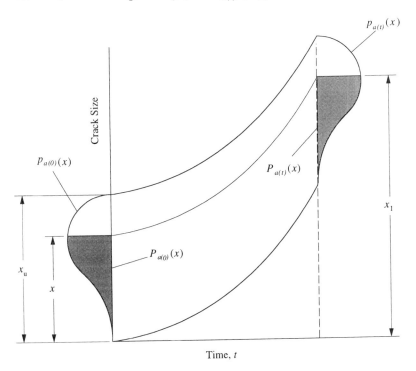

**Fig. 6.3** Equivalent initial flaw size (EIFS)

The procedure described makes use of a single crack growth law to describe the entire growth history. Manning and Yang (1987) have also proposed the use of a procedure in which the crack growth is described by one law for crack lengths up to a given value and a second law for values greater than this value. This two segment approach increases the curve fitting capability of the analysis. The possibility of extensions to more than two segments has also been discussed (Manning and Yang, 1987).

The growth histories for cracks with the same value of equivalent initial flaw size may diverge from one another as shown in Fig. 6.2 for two initital values. Manning and Yang (1987) have, consequently, introduced the use of a stochastic growth rate law of the form

$$\frac{da}{dt} = XQ[a(t)]^b \qquad (6.4)$$

where $X$ is a log normal random variable with a median value of unity. The deterministic growth law of equation (6.2), therefore, represents the median growth behavior. These investigators have also proposed the use of a two segment stochastic growth rate law which is analogous to the deterministic two segment law.

The stochastic growth analysis method has been proposed for use in the prediction of aircraft structural maintenance requirements in terms of risk (Manning and Yang, 1992). It also may be useful in scheduling required fleet structural maintenance actions.

In view of the uncertainties associated with crack nucleation and small crack growth, a statistically based methodology would appear to be sound. The equivalent initial flaw size values and the distribution functions are, however, quite dependent on the specific set of conditions for which they have been derived. Wanhill (1986) has, for example, cited references which indicate that the initial growth of small cracks is strongly influenced by local geometry and stress state, fretting, load history, fastener fit, hole preparation and local microstructure. This observation suggests that appropriate background data must be available for each application. Evidence has, however, been presented which indicates that the concept can be effectively used to evaluate initial quality when attention can be focused on a single type of flaw.

The effect on the initial quality of the metal processing of thick plates has been investigated by Owen, Bucci and Kegarise (1989). Their studies were directed towards an evaluation of the extent to which initial quality could be improved through a more stringent control of processing variables. The test material was the aluminum alloy 7050-T7451 in plate form, and the dominant type of flaw was microporosity. Properties were compared for stocks processed before and after the improved quality control was introduced. Ultrasonic examination of the stocks were conducted and it was found that the porosity distribution was shifted towards smaller values by improved quality control. Also, significant increases in the ductility and strength properties were obtained.

Smooth bar, axial fatigue tests were conducted on three lots of material produced over a period during which process control was progressively improved. Failures in these tests were found to originate at micropores on or just below the specimen surface. Plots of cumulative failure versus fatigue lifetime formed distinct curves for three lots of material. The lot produced first had the shortest fatigue lives and the lives for the third lot were the longest. The smooth bar fatigue tests, therefore, provided a good method for evaluating the effect of initial quality.

Owen, Bucci and Kegarise (1989) also conducted fatigue crack tests and they found that crack growth was not sensitive to differences in microporosity when the crack was much larger than the scale of the microstructure. This indicated that the sensitivity to the size and distribution of micropores was primarily confined to crack nucleation and small crack growth.

Since a characterization of the initial flaws was available, Owen, Bucci and Kegarise (1989) were able to use the equivalent initial flaw size concept to correlate their data. To account for the small crack growth anomaly in the near threshold region, they used the crack growth rate approximation depicted in Fig. 5.5. They found that an equivalent initial flaw based on a semi-elliptical surface crack provided a good correlation with micropore size versus cycles to failure data for $\sigma_{max} =$ 172.4 MPa and $R = 0.1$. Their results indicate that when the features of the dominant initial flaw are clearly characterized, a correlation with a model for an equivalent initial flaw is possible.

Murakami and Endo (1992) have conducted extensive investigations of the effect of small defects and nonmetallic inclusions on fatigue strength, and have introduced a geometrical parameter which appears to be useful in correlating fatigue strengths. The parameter is defined as the square root of the area of a defect projected onto a plane perpendicular to the maximum tensile stress. Although the parameter is empirical in origin, it focuses on defects which often prove to be sites for crack initiation. Identification and characterization of defects in this manner may ultimately provide a basis for establishing another connection with the equivalent initial flaw concept.

## 6.3  MULTIPLE SITE DAMAGE

Since damage tolerance and durability emphasize the capability of an aircraft structure to resist failure which originates from a flaw, it is clear that the existence of flaws is accepted as being unavoidable. The flaw which has been of primary concern in previous sections has been the crack, and attention has focused on the crack growth of a single crack. It is not likely, however, that only a single crack will be present in a complex structure such as an airframe. One can, therefore, anticipate that considerations of damage tolerance and durability should include possible interactions between damage at multiple sites, i.e. 'multiple site damage'.

A prevailing trend toward extending the usage of aircraft beyond their initial service goal has created considerable interest in the structural integrity of 'aging aircraft'. Books and conference proceedings have, in fact, focused on the associated problems (Atluri, Sampath and Tong, 1990; Atluri *et al.* 1992; Blom, 1993b).

Multiple site cracking is common in smooth bar fatigue tests which are conducted to determine stress or strain versus cycles to failure properties. It has also been observed in experiments designed to investigate the fatigue behavior of small cracks. In experiments conducted by Swain (1992) it was, in fact, necessary to introduce criteria for evaluating the validity of small crack test results in which interactions between multiple cracks may have been possible. Such interactions could be expected to complicate the interpretation of test results.

In airframes a common concern involves multiple cracking in riveted joints (Schijve, 1992). Schijve has suggested the complexity of load diffusion can be expected to make the prediction of early crack growth in lap joints difficult. Wanhill (1992) has performed analyses of tensioned sheets with holes which had edge cracks. The cases of one edge crack and two edge cracks of equal lengths were included. Also, both fully remote loading and pin loading plus remote loading conditions were considered. To account for widespread scatter in the small and large crack data, he used two growth rate versus $\Delta K$ representations. One provided a median fit of the data and the other represented an upper bound below which most of the data points were located. Using load spectra for a Fokker 100 at a fastener hole in the cabin and at a wing to fuselage fastener hole, he determined crack growth histories for the cases of remote loading and for remote plus pin loading for a single edge crack and for two edge cracks. The latter represents a simple case of multiple site damage. The NASA/FLAGRO (1989) code was used for the remote loading case and a code described by Bossman and Lof (1985) was used for the remote loading with pin loading case.

Crack growth histories determined by use of the median growth rate representation were acceptable in terms of an economic life of 90000 hours for the remote loading of both the pressurized cabin and fuselage – wing cases. For remote plus pin loaded conditions the fuselage – wing case was acceptable, but the pressurized cabin case was unacceptable. The crack growth histories determined by use of the upper bound growth rate representation generally exceeded economic life requirements with the remote plus pin loading case for the pressurized cabin developing the longest crack lengths.

From his results Wanhill (1992) concluded that crack growth is almost always slower when no load transfer occurs through fasteners, and that mechanically connected joints with high load transfer through fasteners are more susceptible to multiple site cracking. His results did not address the effects of the several complications.

1. Bending which can be present in lap joints.
2. The presence of biaxial loading.
3. Interactions of cracks of unequal length at a fastener hole.
4. Interactions of unequal cracks approaching one another.
5. Hole geometry, fastener fit and residual stresses.
6. Load distribution in a hole and fretting.
7. The effects of combined pin and adhesive joining methods which can dramatically improve joint durability.

These complications and the possible presence of what Lincoln (1985) describes as a 'rogue' defect which might properly be represented by the upper bound solution introduce uncertainties and indicate that the use of probabilistic rather than deterministic methods may ultimately be expedient.

Efforts have been made to develop more comprehensive methods of evaluating the effects of multiple site damage. Park and Atluri (1993) have performed finite element analyses of the growth of multiple cracks near a row of fastener holes in a bonded, riveted lap joint in a pressurized aircraft fuselage. In the numerical trials which they conducted they observed the development of a 'catch-up' phenomenon in which short cracks ultimately begin to grow faster than the cracks which initially were longer.

Beuth and Hutchinson (1994) have conducted a finite element analysis of a cracked lap joint with three rows of rigid pin fasteners. Two types of damage were investigated. For one, cracks of equal length on both sides of the top row of rivet holes were considered. For a second case, long cracks alternated with short cracks along the top row of rivet holes. Both the stress intensity factors for the cracked holes and the stress concentration factors for the uncracked holes were determined. The redistribution of stress and the compliance changes of the joint were determined. It was concluded that though the results indicated a sensitivity of predictions to the details of modeling, further theoretical and experimental work is necessary to better characterize the interactions of the behavior. The results did not, for example, indicate that short cracks would 'catch-up' with long cracks, so no explanation for the relative uniformity of multiple crack lengths observed in both tests and actual aircraft emerged from the analysis.

Park and Atluri (1993) have used finite element analyses to evaluate the residual strength of pressurized fuselage panels with multiple fatigue cracks. They found that the residual strength of a panel in which a main crack is approaching a row of rivet holes is significantly reduced even when the rivet holes do not have edge cracks.

Tong, Greif and Chen (1994) have used the hybrid finite element method to determine the residual strength of aircraft panels with multiple site damage. A panel consisting of a rectangular sheet riveted to three stiffeners was analyzed. The direction of loading was parallel to the stiffener directions. The cases analyzed included a large central crack and a large central crack with two smaller colinear outer cracks. In each case the cracks were centered over a stiffener. The effect of a broken central stiffener was also examined. Stress intensity factors, stress concentration factors, and rivet loads were determined. The method developed provides for a determination of residual strength as a function of various combinations of damage. These include multiple cracking with and without a broken central stiffener. Diagrams which display the residual strength of a panel as a function of crack length for different combinations of damage are presented. From these diagrams conclusions regarding the conditions required for crack arrest can be deduced.

The analyses of Wanhill (1992), Park and Atluri (1993) and Park *et al.* (1993) and are primarily directed towards a determination of the consequences of selected patterns of damage. Such results can, therefore, be useful both in the detailed design of new structural systems, and in modification and repair operations.

## 6.4  BONDED PATCH REPAIRS

The objective of periodic inspections is to determine whether or not unacceptable damage has been incurred since the last inspection. The discovery of such damage can result in the replacement or modification of a component or some form of localized repair. One type of repair which was introduced during the latter part of the 1970s involves the use of composite patches which are adhesively bonded to surfaces containing cracks. These patches have been used to repair a wide variety of cracked structures. These include aircraft wing skins, wing planks, door frames and landing wheels. Techniques for designing patch repairs have, in fact, been the primary topic in a compilation of articles edited by Baker and Jones (1988).

Patch repairs do not simply involve applying a patch to a cracked structure. Certain requirements must be satisfied. The patch must, for example, reduce the stress intensity factor to a value  for which crack growth will either be arrested or have an acceptable rate of growth. Also, to maintain the integrity of the patch, the adhesive shear strains and patch strains must not exceed critical values. Clearly, the satisfaction of these requirements means that a detailed stress analysis of the coupling between the patch and the cracked body must be performed. A variety of methods have been developed for designing bonded patches. These include the use of finite elements (Park, Ogiso and Atluri, 1992), boundary elements (Young, Cartwright and Rooke, 1988) and closed form analytical relations (Rose, 1982). For complex or highly critical repairs a detailed three dimensional finite element analysis is generally recommended. In addition to a modeling of the composite patch and the cracked component to be repaired, a model of the adhesive layer must be developed. The latter should properly account for residual stresses arising from the differential thermal contraction which occurs during cooling from the adhesive cure temperature.

Poole and Young (1992) have conducted an investigation in which the predictions obtained from analysis were compared with experimental measurements of both the stress distribution and fatigue crack growth. The specimen used was a centrally cracked sheet to which patches were bonded to both faces, i.e. a double sided patch repair. Unidirectional carbon fiber reinforced plastic patches were used. The fibers were aligned normal to a 20 mm through crack and loads were applied normal to the crack. The specimen width was 108 mm. Five sheet thicknesses ranging from 1.55 to 24.8 mm were investigated and the patch thicknesses were chosen to maintain a constant ratio of patch stiffness to sheet stiffness. This resulted in a patch thickness to sheet thickness ratio of 0.1645. This choice was based on experience with previous patch repair designs.

The cracked sheet was modeled by use of the boundary element method. The composite patch was modeled using mixed stress – displacement finite elements. The adhesive forces developed between the patch and the sheet were treated as internal body forces in the sheet.

The values of the stress intensity factors were largest at the mid-point of the crack front and decreased as the front approached the sheet surface. For the same applied stress the stress intensity factors decreased with decreasing thickness. This thickness effect is not observed for the unpatched specimens, i.e. the values of the stress intensity factors are essentially independent of thickness.

The objective of the patch repair is to reduce the stress intensity factor. The calculated reductions for all sheet thicknesses were substantial, but the largest was for the thinnest sheet. The reductions ranged from 77% for the 1.55 mm sheet to 50% for the 24.8 mm sheet.

Poole and Young (1992) also used simplified two dimensional and one dimensional analyses for the same test cases and found that the simpler approximations were good up to a sheet thickness of 3.1 mm. Beyond that thickness, the approximations overestimated the reduction in stress intensity factor, i.e. they were nonconservative.

Patch surface stress distributions for test specimens were determined by the use of strain gages and a thermal emission technique. These results were compared with the distributions obtained from the three dimensional model analysis. The results were found to be in good agreement. Also, values of stress intensity factor range inferred from fatigue crack tests on patched specimens were compared with theoretical values. Again, the agreement was good.

Park, Ogiso and Atluri (1992) have developed finite element methods for analyzing a variety of patch repair problems. These include the cracked plate and a plate with a crack at the edge of a hole. The solution procedure for the latter problem can also be applied to configuations involving holes in the legs of stiffeners. The programs which they have generated are versatile and can be adapted to perform parametric studies of the effects of geometry, loading conditions, adhesives and patch design.

# 7

# Structural integrity of polymeric matrix composite laminates

## 7.1 INTRODUCTION

Numerous possibilities exist in combining constituents to form a composite. Among these are the broad groups of metal-matrix, ceramic-matrix and polymeric-matrix composites which were discussed in Chapter 3. Of these, the polymeric-matrix composites have had the most wide acceptance in engineering applications. Applications have included short fiber composites, woven fabric composites and long and continuous fiber composites. Current primary structural applications, mainly in the aerospace industry, are in the form of laminates with long and continuous fibers in a fixed direction within each layer. Fatigue damage and low-velocity impact damage are two of the main concerns for ensuring the safety and the long-term integrity of such components as composite rotor blades and composite panels. The loading and lay-up construction can become quite complex in applications such as rotor hubs, flexbeams and built-up roof structures.

The equivalent of a macro-crack in a metal is, in a composite laminate, a delamination, i.e. an interlayer separation. Internal delaminations are typically preceded by an accumulation of matrix cracks. Edge delaminations occur due to interlaminar normal or shear stress concentrations in certain stacking sequences. Ryder and Crossman (1983) performed fatigue tests on a wide range of unnotched, tensile loaded graphite/epoxy laminates. In $[0]_4$ specimens the strain at failure was 10 to 20% less than that for monotonic loading. 0° fiber fractures were observed before failure, and there was a large scatter in the fatigue life data.

In symmetric layup specimens with $[0/\pm45]_s$ the fatigue strain at failure was always 15 to 25% below monotonic strain to failure. Matrix crack spacing saturation was observed for both the +45° and –45° plies. Cracks often did not cross the coupon width until as much as 50% of the fatigue life had been consumed (the average crack spacing was 0.5 mm in the +45° plies and 0.8 mm in the –45° plies). Delamination occurred primarily at the +45/–45 interfaces and secondarily at the 0/+45 interfaces. The +45/–45 interface delamination was confined to a narrow edge region. Delamination extension in the coupon width direction occurred prior to transverse matrix crack saturation. The final fracture displayed 0° fibers fractured along the +45° direction and stiffness loss was up to 7% . Some 0° fiber breaks were observed prior to coupon failure.

In [0/90/±45]$_s$ layups, the matrix crack spacing in the 90°, +45° and in the –45° plies saturated in less than 40% of the fatigue life and led to a 2 to 3% decrease in stiffness. Delaminations started at free edges after transverse crack saturation, and the 90/±45 interface was the dominant location. There was regular and continuous stiffness loss of up to 18% and no obvious 0° fiber fractures were observed prior to coupon failure. Again, the strain at failure was always 15 to 25% below the monotonic strain to failure and the 0° plies fractured along the +45° direction.

In [0$_2$/90$_4$]$_s$ layups, the transverse matrix cracking saturated in many regions early in the fatigue life (5000 cycles or less), and 0° longitudinal splits occurred. Moreover, 0/90 interface delamination grew from the intersection of the 0° splits and the 90°matrix cracks and 0° fiber fracture occurred in large groups. There was a stiffness loss of up to 17% prior to failure, and the strain at failure was 20 to 30% below the average strain to failure of monotonically loaded coupons. At failure, the coupons essentially exploded. The 0° plies again fractured along a 90° direction.

Some of the experimental results obtained by Ryder and Crossman (1983) are presented in Fig. 7.1. In Fig. 7.1 (a) the trend of the stress versus cycles to failure curves would, if drawn, be almost horizontal. By contrast, the data points of Fig. 7.1 (b) exhibit significant downward trends with increasing cycles to failure. A comparison of the layups for the specimens exhibiting these two distinct trends reveals that the primary difference is the inclusion of 90° plies for the specimen data shown in Fig. 7. 1 (b). The stiffness of the 0° and the 90° plies represents extremes for loading in the 0° direction, and this incompatibility leads to the development of the failure mechanisms cited above for the layups containing the 90° plies.

The results of Ryder and Crossman (1983) reveal how fatigue damage mechanisms are dependent on stacking sequence. It should, however, be noted that whereas an absence of 90° plies is beneficial for tensile loading in 0° direction, it would not follow that it would be advantageous for applied biaxial tensile fields.

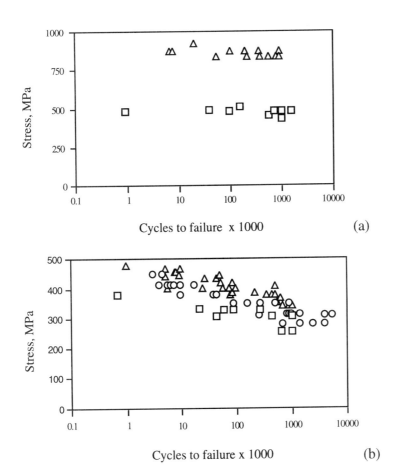

**Fig. 7.1** Stress versus cycles to failure for (a) [0/±45]$_s$ □ and [0/45/0$_2$/–45/0]$_s$ △ and (b) [0$_2$/90$_4$]$_s$ □ , [0/45/90/–45]$_{2s}$ ○ and [0/90/±45]$_s$ △ (Ryder and Crossman, 1983)

## 7.2 STRAIN CONCENTRATION DUE TO EDGE AND INTERNAL DELAMINATIONS

To illustrate the development of strain concentrations due to delaminations consider a graphite/epoxy $[\pm 25/90_n]_s$ laminate for which $2n$ is the total number of 90° lamina and $n$ is taken to vary from 1/2 to 8. This is chosen as a representative laminate because Crossman and Wang (1982) have reported detailed test results. O'Brien (1985) has provided formulae and procedures that account for strain concentration effects. Edge delaminations typically form at interfaces between 90° plies and adjacent plies. Measurements of reduction in stiffness from the initial value, $E_L$, agreed well with $E^*$ where

$$E^* = \frac{1}{t}\sum_{i=1}^{m+1} E_i t_i,$$ (7.1)

and $m + 1$ is the number of sublaminates formed by the delamination, $E_i$ is the modulus of the $i$th sublaminate formed by the delamination and $t_i$ is the thickness of the $i$th sublaminate. The $E_i$ are calculated from laminated plate theory. The difference in $E_L$ and $E^*$ reflects the loss of transverse constraint in the sublaminates formed by the delamination. As an example, for the $[\pm 25/90_n]_s$ family of layups with delaminations in the $-25/90$ interfaces,

$$E^* = \frac{4E(\pm 25)_s + 2nE_{90}}{(4 + 2n)}.$$ (7.2)

Fig. 7.2 shows through-thickness free-body diagrams for $[\pm 25/90_n]_s$ laminates with (a) edge delamination in the $-25/90$ interfaces, (b) localized, internal delamination in the $-25/90$ interfaces growing from a $90_n^\circ$ matrix ply crack, and (c) both edge and localized, internal delaminations. The local strain concentration, $K$ from the top (T) to the bottom (B) through-thickness cross-section for this case is

$$K = \frac{\varepsilon_B}{\varepsilon_T} = \frac{E_T t_T}{E_B t_B}$$ (7.3)

where $\varepsilon_B$ and $\varepsilon_T$ are the strains at the bottom and top, respectively. For case (a), where only edge delamination occurs, the moduli of the top and bottom sections are identical, and are equal to $E^*$, as calculated by equation (7.2), and the thicknesses of the top and bottom cross-sections are identical. Hence, edge delamination does not result in local strain concentrations in the individual plies.

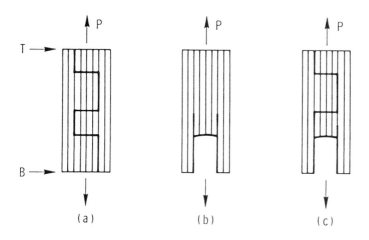

**Fig. 7.2** Free-body diagrams for progressive delamination in a $[25/90_n]_s$ laminate (from O'Brien, 1985)

For case (b), however, the modulus and the thickness of the top cross-section are those of the original laminate. The modulus of the bottom cross-section is simply the modulus of a $[\pm 25]_s$ laminate. The thickness of the bottom cross-section is simply the thickness of the four $\pm 25°$ plies, because the isolated 90° plies no longer carry load. Hence, the isolated 90° plies are modeled as if they have been removed from the laminate. In case (c), a combination of edge and local delamination occurs. In this case, the modulus on the top cross-section is equal to $E^*$ from equation

(7.2) and the thickness is the original laminate thickness. The modulus and thickness of the bottom cross-section are simply the modulus and thickness of a $[\pm 25]_s$ laminate. Thus the resulting strain concentrations, $K$, in the $\pm 25°$, load bearing plies can be calculated for different layups, e.g. $n = 4, 6, 8$. This computational procedure can be illustrated by solving problem 7.2.

## 7.3 INFLUENCE OF DAMAGE ON LAMINATE STIFFNESS

As matrix cracks accumulate, and as delaminations form and grow, the stiffness of a laminate decreases. The amount of stiffness loss associated with matrix cracking depends upon the fiber orientation of the cracked ply, the laminate layup, the relative moduli of the fiber and the matrix, and the crack spacing or density of cracks in the ply. For example, Caslini, Zanotti and O'Brien (1987) derived an equation for stiffness loss due to matrix cracking in the $90°$ plies of cross-ply laminates (the $0°$ ply is the uncracked ply) of the form

$$E = \frac{E_L}{1 + \left(\frac{1}{\lambda s}\right)\left(\frac{c}{d}\right)\left(\frac{E_{22}}{E_{11}}\right)\tanh(\lambda s)} \ , \tag{7.4}$$

where

$$\lambda = \left(\frac{3G_{12}(c+d)E_L}{c^2 dE_{11}E_{22}}\right)^{1/2} . \tag{7.5}$$

$E$ and $E_L$ are the moduli of the cracked and the uncracked laminate, respectively, $E_{11}$, $E_{22}$ and $G_{12}$ are the laminate longitudinal, transverse and shear moduli, respectively, $s$ is the crack half-spacing, and $c$ and $d$ are the thicknesses of the cracked and the uncracked plies, respectively. As the crack density increases, or equivalently, as the crack spacing, $2s$, decreases, the stiffness of the laminate will decrease.

The amount of stiffness loss due to delamination can be estimated by using equation (7.1) to find $E^*$, the modulus of the delaminated laminate. There is also a dependence on the location and the extent of the delamination. As delaminations form and grow in a given interface, the laminate stiffness decreases as the delamination size, $a$, increases.

O'Brien (1982) has given a simple relation for the stiffness loss associated with edge delamination, i.e.

$$E = \frac{\left(E^* - E_L\right)a}{b} + E_L \tag{7.6}$$

where $a/b$ is the ratio of the delamination size to the laminate half-width.

Internal delaminations initiating from matrix cracks will affect laminate stiffness differently from edge delaminations. O'Brien (1985) has derived an equation for the stiffness loss associated with delaminations initiated from matrix cracks. It has the form

$$E = \left[\left(\frac{a}{l}\right)t_L\left(\frac{1}{t_D E_D} - \frac{1}{t_L E_L}\right)\right]^{-1} \tag{7.7}$$

where $a/l$ is the ratio of the delamination length to the laminate length, and $E_D$ and $t_D$ represent the modulus and the thickness of the locally delaminated region in the vicinity of the matrix crack. In Fig. 7.3 the composite gauge length, $l$, is divided into a 'locally delaminated' region, $a$, and a 'laminated region', $(l - a)$. The $t_D$ is the thickness of the locally delaminated region that carries the load, i.e. the thickness of the uncracked plies $t_D = t_1 + t_2$. The locally delaminated modulus, $E_D$, is calculated using laminated plate theory, and it is analogous to $E^*$ calculated for edge delaminations for which the delaminated plies are isolated as shown in Fig. 7.3.

## 7.4 DELAMINATION INITIATION

Cyclic loading can introduce local delaminations which are characterized by a maximum energy release rate, $G_{max}$. A plot of $G_{max}$ versus $N$ is similar to an $S - N$ diagram. Data from several materials with brittle and tough matrices indicate that below $10^6$ cycles, the maximum cyclic value of $G$ that causes initiation of delamination at $N$ cycles is

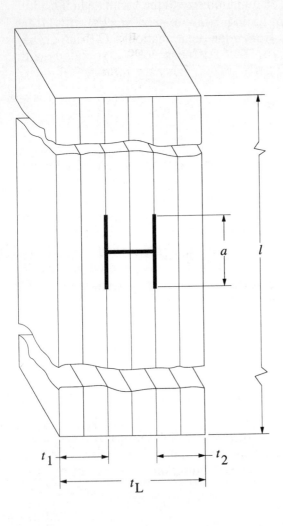

**Fig. 7.3**   Internal delaminations initiating from a matrix crack

$$G_{max} = m \log N + G_c, \qquad (7.8)$$

where $G_c$ characterizes the initiation under static loading and $m$ is a material parameter. The plot shown in Fig. 7.4 (O'Brien, 1990) indicates the variation in $G_{max}$ for four composites.

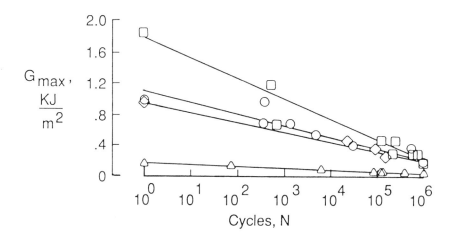

**Fig. 7.4** $G_{max}$ versus log $N$ plots: AS4/PEEK □ ; AS6/HS7 ○ ; C6000/H205 ◇ ; T300/5208 △ (from O'Brien, 1990)

O'Brien (1985) has proposed an equation for the strain energy release rate associated with local delaminations which are initiated at matrix cracks when the composite is subjected to a nominal axial stress $\sigma$. It has the form

$$G = \frac{\sigma^2 t_L^2}{2} \left( \frac{1}{t_D E_D} - \frac{1}{t_L E_L} \right) \qquad (7.9)$$

where $t_L$ is the laminate thickness, $E_D$ is the modulus of a locally delaminated cross-section and $t_D$ is the thickness of a locally delaminated cross-section that carries load, i.e. the thickness of the uncracked plies. Using the foregoing equation provides a method for determining the number of cycles for the initiation of the first local delamination, $N_1$.

$$\log N_1 = \frac{1}{m}\left[\frac{\sigma_{max}^2}{2}t_L^2\left(\frac{1}{t_D E_D} - \frac{1}{t_L E_L}\right) - G_c\right] \qquad (7.10)$$

where $G_c$ is the critical value. Although the toughness of the matrix has a very strong effect on $G_c$, it has been found that it has very little influence on delamination initiation at $10^6$ cycles. Therefore, the slope, $m$, as measured by fitting the delamination initiation data to equation (7.8), will be lower for a brittle matrix composite than for a tough matrix composite.

As shown in Fig. 7.5, the thickness and modulus terms in equation (7.9) change for each successive local delamination that forms through the thickness. The values of $t_D$ and $E_D$ or a 45/–45 local delamination in a $(45/–45/0)_s$ laminate then become the $t_L$ and $E_L$ values used for the next local delamination that forms through the thickness. Note that the $t_D$ and $E_D$ correspond to $t_{LD}$ and $E_{LD}$ in Fig. 7.5. Also, $t_L$ and $E_L$ correspond to $t_{LAM}$ and $E_{LAM}$. Therefore, as local delaminations accumulate through the thickness under a constant $\sigma_{max}$, the driving force, $G$, for each new delamination changes. This is essentially an 'accumulative damage' approach. To calculate the number of cycles, $N_i$, for each successive local delamination to form, the initiation criterion of equation (7.8) can be used with the $G$ expression of equation (7.9) to give

$$\log N_i = \frac{1}{m}\left[\frac{\sigma_{max}^2}{2}(t_L^2)_i\left(\frac{1}{t_D E_D} - \frac{1}{t_L E_L}\right)_i - G_c\right]. \qquad (7.11)$$

Fatigue failures under fixed cyclic load conditions occur typically after the global strain has increased because of the fatigue damage growth, but before the global strain reaches the static failure strain, $\varepsilon_F$. In fact, once delaminations initiate at matrix ply cracks anywhere through the laminate thickness, the local strain increases significantly throughout the remaining through-thickness cross-section.

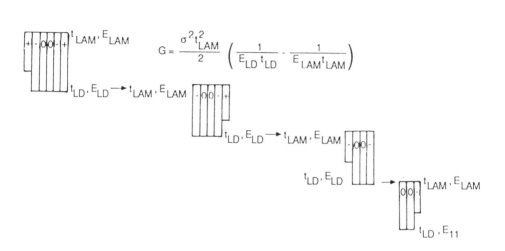

$$G = \frac{\sigma^2 t^2_{LAM}}{2}\left(\frac{1}{E_{LD}\,t_{LD}} - \frac{1}{E_{LAM}\,t_{LAM}}\right)$$

**Fig. 7.5** Representations of progressive through-thickness delamination (from O'Brien, 1990)

A method for accounting for this mechanism has been suggested by O'Brien (1990). Each time a delamination initiates from a matrix crack, the local strain in the remaining through-thickness cross-section increases. The new local strain is equal to a strain concentration factor $K_\varepsilon$ times the global cyclic strain, $\varepsilon_{max}$. As shown in Fig. 7.6 (a), the local strain increases until it reaches the static failure strain, $\varepsilon_F$. The effect of the decreasing static failure strain on the global strain is shown in Fig. 7.6 (b). The strain concentration factor proposed by O'Brien is

$$K_\varepsilon = \frac{E_L t_L}{E_D t_D}. \qquad (7.12)$$

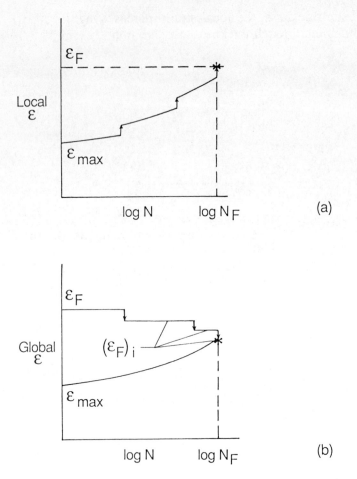

**Fig. 7.6** Changes in local and global strains with progressive delamination (from O'Brien, 1990).

Notice that this approach does not require a prediction of damage growth with fatigue cycles if the laminate stiffness loss, and hence the increase in global strain, can be monitored in real time. When this is possible, only the incremental decreases in the effective $\varepsilon_F$ need to be predicted to determine fatigue life. This may be accomplished by assuming that matrix cracks exist in all of the off-axis plies. This assumption is analogous to assuming the existence in a metal of the smallest flaw that can be detected non-destructively to assess damage tolerance. The resulting fatigue life, $N_F$, will be

$$N_{\mathrm{F}} = \sum_{i=1}^{p} N_i, \qquad\qquad (7.13)$$

where $p$ is the number of local delaminations that form through the thickness of the laminate before failure.

## 7.5 ON MODE-DEPENDENT CYCLIC DELAMINATION GROWTH LAWS

Several investigators have analyzed the strain energy release rate, $G$, with respect to predicting edge or internal delamination growth. In these analyses, a Griffith-type fracture criterion is employed, and it is assumed that whether or not further delamination occurs depends on the magnitude of the fracture energy, which is defined as the energy required to produce a unit area of new delamination. In mixed mode linear elastic fracture mechanics this quantity can be expressed in terms of the mode I and II stress intensity factors, $K_{\mathrm{I}}$ and $K_{\mathrm{II}}$, respectively (Hellan, 1984). When more than one mode is active at a crack tip, it is reasonable to expect that the crack extension process will be dependent on the relative loading intensities of the component modes. This expectation has prompted the introduction of 'mixity' parameters which are functions of ratios of the component stress intensity factors. These parameters can, of course, vary during the course of mixed mode crack extension.

Unlike the growth of macro-cracks in metals, where the crack can change its growth direction and transform an initially mixed mode condition into a mode I loading condition, delamination growth is restricted to extend along a fixed direction (the layer interface). For example, under compression loading, which causes buckling of the delaminated layer, it has been shown that during the initial postbuckling phase, the mode mixity changes with the applied strain, and it depends on the relative delamination thickness $h/T$. For example, a study by Kardomateas (1993) has shown that a higher mode I component is present with delaminations further away from the surface. The mode mixity also changes as the delamination grows, with an increasing mode II component under a constant applied compressive strain.

For cyclic loading that causes a variation of the strain energy release rate from $G_{\mathrm{min}}$ to $G_{\mathrm{max}}$, the stresses near the tip of the delamination are completely described by $G_{\mathrm{max}}$, the load ratio $\alpha = G_{\mathrm{min}}/G_{\mathrm{max}}$ which expresses the ratio of minimum to maximum loading, and the mixity parameter, $\psi = \tan^{-1}(K_{II}/K_I)$ which expresses the relative amounts of mode I (opening) and mode II (shearing) components. This follows because these three parameters describe both the intensity and the variation of the stress field near the delamination tip for a given geometric configuration and loading.

During a cycle of loading, the stresses and strains near the tip of a delamination are completely specified by $G_{max}$, $\alpha$ and $\psi$, so it can reasonably be assumed that any phenomenon occurring in this region is in turn controlled by these parameters. The amount of crack extension per cycle of loading is just such a phenomenon. In functional form

$$\frac{da}{dN} = f(G_{max}, \alpha, \psi). \tag{7.14}$$

The mode dependence of the delamination growth process is not yet fully understood and is a subject of considerable research. Chai (1992) has conducted an experimental evaluation of mixed-mode fracture in adhesive bonds in order to elucidate the effect of the mode of loading on the interlaminar fracture toughness. The mixed-mode fracture behavior was characterized by essentially two bond thickness regimes. In the first, which is limited to a few micrometers, mode interaction occurred promptly upon the application of any combination of loads. In the second regime (thickness greater than about five micrometers), mode interaction occurred only when the shearing component of the energy release rate exceeded a certain value. In the case of the ductile BP-907 adhesive, that value was approximately 55% of the pure mode II toughness, $G_{II}^c$. Below this value of $G_{II}$, the fracture toughness was that for pure mode I, $G_I^c$. Test methods for determining the mode I, $G_I^c$, and mode II, $G_{II}^c$ fracture toughness values for different composite systems have been reviewed by Sela and Ishai (1989). Of these, the most widely used are the double cantilever beam (DCB) test for $G_I^c$ and the end notched flexural (ENF) test for $G_{II}^c$.

Table 7.1 from Russell and Street (1987) presents values of fracture toughness for some common composites. The data presented are mean values with the indicated standard deviations. These composites include a first generation high temperature graphite/epoxy (Hercules AS1/3501-6), a second generation high temperature graphite/epoxy (Hercules AS4/2220-3), an intermediate temperature, rubber toughened graphite/epoxy (Hexcel C6000/F155) and a high temperature graphite/thermoplastic (ICI VICTREX APC2 which is AS4 fiber/PEEK composite). The $G_I^c$ values listed in Table 7.1 refer to the fracture energies for the onset of delamination. The $G_I^s$ values refer to the stabilized or plateau fracture energies which were measured after the full development of fiber bridging. The range of values for the four composites is substantial.

Table 7.1  Fracture toughness of common composites (Russell and Street, 1987)

| Material | Fiber volume fraction | $G_I^c$ $(J/m^2)$ | $G_I^s$ $(J/m^2)$ | $G_{II}^c$ $(J/m^2)$ |
|---|---|---|---|---|
| AS1/3501-6 | 0.62 | 110±5 | 190±10 | 605±30 |
| AS4/2220-3 | 0.61 | 160±10 | 255±15 | 750±25 |
| C6000/F155 | 0.51 | 495±25 | 510±20 | 900±50 |
| AS4/1PC2 | 0.66 | 1330±85 | 1540±60 | 1765±24 |

Because interlaminar, resin-rich interface layers in laminated composites are typically several micrometers thick, it would be expected that mixed-mode, interlaminar fracture would be characterized by some degree of prompt interaction between the modes. In a related study on glass/epoxy, Liechti and Chai (1992) measured the toughness of the glass/epozy interface over a wide range of mode mixities and found a toughening effect associated with increasing in-plane shear components. Optical interference measurements of the normal crack opening displacement revealed large variations in plastic zone size with changes in mode mixity. The plastic zone sizes followed the same trends that the toughness exhibited with mode mixity. Although small scale yielding was observed for all cases, there were large changes in size as the shear component increased.

From the above discussion it may be concluded that the toughness $\Gamma_0$ depends on the mode mixity, $\psi$, and that $\Gamma_0(\psi)$ increases with increasing $|\psi|$ (increasing mode II component). A simple, one parameter fracture criterion for mixed mode loading has been proposed by Hutchinson and Suo (1992). It has the form

$$\Gamma_o(\psi) = G_I^c \left[ 1 + (\lambda - 1) \sin^2 \psi \right]^{-1}.$$ (7.15)

The parameter $\lambda$ adjusts the influence of the mode II contribution in the criterion, and it must be determined experimentally by obtaining mode-interaction curves (Chai, 1992). For pure mode I, $\psi = 0$, and $\Gamma_0 = G_I^c$. For pure mode II, $\psi = 90°$, and $\Gamma_0 = G_{II}^c$. Equation (7.15) indicates that

$$\lambda = G_I^c / G_{II}^c$$ (7.16)

where $G_I^c$, $G_{II}^c$ are the values of the pure mode I and pure mode II toughness, respectively. For graphite/epoxy, a value of $\lambda = 0.30$ is typical. The limit $\lambda = 1$ is the case of the classical mode-independent toughness, i.e., $\Gamma_0 = G_I^c$ for all mode combinations.

From the preceding results, the effects of mode-dependent toughness on the growth characteristics can be described by defining

$$\tilde{G} = \frac{G}{\Gamma_0(\psi)} = \tilde{G}(\varepsilon_0, \psi) .$$ (7.17)

This definition provides a mode-adjusted crack driving force in the sense that the criterion for crack advance is $G/\Gamma_0(\psi) = 1$.

For stable delamination growth Russell and Street (1987, 1988) have examined a form of equation (7.14) which resembles the Paris law. The growth rate equation considered was

$$\frac{da}{dN} = C \left( \Delta \tilde{G} \right)^m,$$ (7.18)

where the range of energy release rate

$$\Delta \tilde{G} = \tilde{G}_{max} - \tilde{G}_{min}.$$ (7.19)

The dependencies of $C$ and $m$ on mixity would have to be deduced from experimental measurements of delamination growth.

Table 7.2 gives measured fatigue delamination growth data for six composites. The coefficient $C$ and the exponent $m$ have been taken as constants in equation (7.18). $G^c$ in Table 7.2 refers to the critical energy release rate in the corresponding mode, i.e. $G_I^c$ or $G_{II}^c$. The first three tests entries involved mode I loading. The remaining eight entries were for mode II loading. The second mode I test for AS1/3501-6 was conducted at 100 °C under dry conditions. All of the remaining tests were conducted at 20 °C under dry conditions.

Russell and Street (1988) found that mode I fiber bridging effects could reduce delamination growth rates by more than an order of magnitude. Delamination behavior is dependant on fabrication procedures which include the geometry of laminate layup. Although it is unwise to rely on fiber bridging for design purposes, it can provide a margin of safety. The effect of elevated temperatures was revealed by a comparison of growth rates for the two test temperatures for AS1/3501-6. The mode I delamination growth rates for fatigue tests conducted at 100 °C were an order of magnitude greater than those at 20 °C. The data in Table 7.2 are based on initial values of fatigue crack growth rate for which fiber bridging effects are minimal.

Table 7.2 Fatigue growth of delamination (Russell and Street, 1987, 1988)

| Material | Test conditions | $G^c$ (J/m$^2$) | $C$ (mm/cycle) | $m$ |
|----------|-----------------|------------------|-----------------|-----|
| AS1/3501-6 | 2.5 Hz, $R = 0.05$ | 110 | 0.0325 | 9.4 |
| AS1/3501-6 | 2.5 Hz, $R = 0.05$ | 110 | 0.0206 | 6.8 |
| AS4/PEEK | 1.0 Hz, $R = 0.05$ | 2100 | 0.0041 | 3.0 |
| AS1/3501-6 | 2.0 Hz, $R = 0$ | 605 | 0.285 | 5.79 |
| AS1/3501-6 | 1.0 Hz, $R = -1$ | 605 | 0.0588 | 3.87 |
| AS4/2220-3 | 2.0 Hz, $R = 0$ | 750 | 0.423 | 5.71 |
| AS4/2220-3 | 1.0 Hz, $R = -1$ | 750 | 0.723 | 3.53 |
| C6000/F155 | 2.0 Hz, $R = 0$ | 900 | 0.196 | 4.52 |
| C6000/F155 | 1.0 Hz, $R = -1$ | 900 | 0.497 | 2.80 |
| AS4/APC2 | 2.0 Hz, $R = 0$ | 1765 | 0.323 | 3.88 |
| AS4/APC2 | 1.0 Hz, $R = -1$ | 1765 | 0.317 | 2.02 |

The last eight entries in Table 7.2 are for mode II loading. For both brittle and toughened matrix materials, interlaminar fatigue growth rates for $R = -1$ are greater than for $R = 0$. This difference was most significant for small values of $\Delta G_{II}$.

By an examination of their test data, however, they observed that the parameters $C$ and $m$ were mode dependent, i.e. they were not constants. They observed that the growth rate $da/dN$, under pure mode I was less than expected on the basis of a mode-independent assumption for $C$ and $m$. This was partially attributed to fiber bridging which occurs in mode I.

If a law of the form of equation (7.18) is to be used, it is clear that the mode dependencies of $C$ and $m$ must be established. Equation (7.15) might be used as a guide in the formulation of the functions $C(\psi)$ and $m(\psi)$. Kardomateas, Pelegri and Malik,1995 proposed the use of

$$m(\psi) = m_I\left[1 + (\mu - 1)\sin^2\psi\right] \qquad (7.20)$$

For pure mode I, $\psi = 0$ and $m = m_I$. For pure mode II, $\psi = 90°$ and $m = m_{II}$ if

$$\mu = m_{II} / m_I. \qquad (7.21)$$

In a similar fashion, the parameter $C(\psi)$ can be written as

$$C(\psi) = C_I\left[1 + (\kappa - 1)\sin^2\psi\right]. \qquad (7.22)$$

For pure mode I, $\psi = 0$ and $C = C_I$. For pure mode II, $\psi = 90°$ and $C = C_{II}$ if

$$\kappa = C_{II} / C_I. \qquad (7.23)$$

The dependencies of $C$ and $m$ on mixity have to be deduced from experimental measurements of delamination growth.

Kardomateas, Pelegri and Malik (1995) have used this mode-dependent fatigue growth law in conjunction with a delamination postbuckling model to analyze delamination growth under cyclic compressive loading. The model is based on the concept of a critical fracture energy for mixed mode loading and is formulated in terms of the energy release rate during stable delamination growth. The proposed model is used as a basis for determining the range of energy release rate in equation (7.18). The values of $m_I$ and $m_{II}$ in equation (7.20) and $C_I$ and $C_{II}$ in equation (7.22) have been deduced from growth data from a cyclic compression test on a unidirectional graphite/epoxy specimen (Kardomateas, Pelegri and Malik,1995). It is shown that an application of a growth law which assumes mode independence fails to provide an adequate correlation of test data. This indicates that if a growth rate law of the form of equation (7.18) is used, mode dependence must be incorporated in the analysis.

## 7.6 DESIGN IMPLICATIONS

The method described by O'Brien (1985) for tracking fatigue in composite laminates is based on the premise that the effect of damage can be assessed by focusing on local strain concentrations which can be determined by a consideration of prevalent delamination modes. Ryder and Crossman (1983) have demonstrated that analyses of this type can be used to correlate stress versus cycles to failure data. This type of approach can be expected to contribute to design decisions.

The problem of stable delamination growth is analogous to the problem of stable fatigue crack growth in metals. More specifically, however, the comparison should be for mixed mode growth. Mixed mode fatigue crack growth in metals is a complex phenomenon. In section 5.6 it is suggested that Pook's proposal (1989) for using the mixed mode fatigue threshold as a limiting condition for design is sound. This suggestion is based on the recognition that the features of mixed mode fatigue cracks in service components can deviate significantly from those assumed in predictive codes. Three dimensional crack surfaces propagating under mixed mode loading can, for example, be curved rather than planar.

Mixed mode delamination growth in laminated composites is also a complex phenomenon. Special tests can be used to evaluate parameters in growth rate equations, and the results can contribute insight into the mechanics of cumulative damage. They can also be useful in the evaluation of alternative fabrication procedures. Ultimately, however, the adoption of composite laminates in critical applications must be based on an evaluation of results from tests on built-up specimens, i.e. evaluations based on tests which duplicate the features of the structure they replace. The test programs conducted by Weller and Singer (1990), (described in section 3.6.2) and by Schijve (1993), (described in section 5.9) are examples which illustrate how this approach provides information required for design.

# 8

# Biomaterials

## 8.1 INTRODUCTION

Advances in medical science have not been dependent solely on contributions from practicing physicians. Individuals usually recognized for their expertise in other disciplines have often been contributors. In the 17th century, for example, the English physicists Robert Boyle and Robert Hooke did pioneering work in the physiology of respiration. The French mathematician and philosopher René Descartes performed anatomical dissections and investigated the anatomy and mechanism of vision. In the 19th century the German physicist Hermann Ludwig Ferdinand von Helmholtz invented the ophthalmoscope and the ophthalmometer, and studied the speed of nerve impulses and reflexive mechanisms. Also in the 19th century the German physicist Wilhelm Conrad Roentgen, by the discovery of X-rays, paved the way for the development of a diagnostic instrument which is commonly used by physicians and dentists. Significant contributions made by other specialists in the fields of chemistry, physiology, bacteriology and microbiology have also been numerous.

During the 20th century the use of surgery to correct deformities and defects by the use of implants has become quite common. Although body materials such as bone and connective tissue are distinct from such implant materials as metal and plastics, the term biomaterials has been adopted to describe both categories. When in place, the interactions between the natural and implanted materials requires that both the physiological and mechanical properties be known.

Implants include the use of pins, screws and plates for broken bones, tiny plastic tubes or stainless steel wires which are used to replace a small bone – the stapes – that transmits sound in the inner ear, and synthetic arteries. In addition to transplanting organs, surgeons are also replacing heart valves and reconstructing knee and hip joints.

From this list of implants it is clear that there is a role to be played in the developing technology by specialists in both materials science and mechanics. Since the primary topic of interest here is fatigue, attention is focused on implanted prostheses which are subjected to intermittent and varying loads. This excludes, for example, a plate covering a hole in a skull, but it includes such devices as heart valves and joint implants. The latter two are considered here as examples which illustrate the exposure of prostheses to fatigue. In view of the consequences of failure of these devices, it is clear that structural integrity is an issue.

## 8.2 A PROSTHETIC HEART VALVE

An average heart beats approximately 30 million times a year. An artificial heart valve intended to last for about 20 years will, therefore, be exposed to fatigue loading of the order of a billion cycles. One heart valve prosthesis which has become widely used is formed by coating a graphite substrate with a thin layer of pyrolytic carbon. The valve consists of a collar with two attached semicircular plates which rotate to open and close the valve. The surfaces are highly polished and have been found to provide a blood clot-free environment (Bokros, 1977). The compatibility of this material with blood and tissue along with its chemical inertness have resulted in its widespread use in prosthetic heart valves.

As noted in section 5.8.2, cyclic fatigue crack growth has been observed to occur in ceramic materials. In view of this behavior Ritchie and Dauskardt (1990) conducted experiments designed to determine the fatigue crack growth behavior of the pyrolytic carbon coated graphite material used in prosthetic heart valves. Compact tension specimens were formed by coating a graphite substrate with a thin layer of a silicon carbide/isotropic pyrolytic carbon alloy. Crack growth tests under a load shedding procedure were performed in order to obtain crack growth rate versus range of stress intensity factor data. A load ratio of 0.1 was used. Tests in both air and a solution which simulated the environment of a prosthesis implant were conducted. Log – log plots of the crack growth behavior exhibited considerable scatter, and little difference was observed between test results in air and in solution. A threshold $\Delta K$ was selected as a value below which crack growth rates did not exceed $10^{-11}$ m per cycle. The value thus obtained was 0.7 MPa$\sqrt{m}$ and this was approximately 50% of the mean of the values of critical stress intensity factors which were obtained.

In fitting the data to a Paris type law the authors determined the value of the exponent to be 18.5. It follows that a change of 10% in the range of the stress intensity factor could result in about a six-fold change in the rate of crack growth. This result is similar to that observed in ceramic materials and it indicates that the test material is not damage tolerant. Since implanted heart valves are not accessible for scheduled inspections, it may be concluded that the permissible values of stress intensity factor range should be well below the threshold value. It should be noted that very stringent inspection procedures are employed to detect flaws prior to implantations. Nevertheless, Ritchie and Dauskardt (1990) maintain that cyclic load behavior rather than constant load data should be considered when estimates of lifetimes are made for implanted heart valves.

## 8.3 PROSTHETIC HIP JOINTS

The second example of fatigue in a prosthesis involves the replacement of a hip joint. Surgical procedures for using metal prostheses to replace damaged and arthritic joints have been developed to the degree that the operation has become quite commonplace. For the hip joint prosthesis the components involved are the femur bone, a metal stem and a polymer cement which bonds the stem to the bone. These components are shown in Fig. 8.1. The mechanical properties of each of these components must be known in order to perform an analysis of the stresses which are developed during various forms of activity. Naturally, the details of the loads which are encountered, i.e. magnitudes, directions, time histories must also be specified.

Early studies of fatigue fractures in bone have been described by Morris and Blickenstaff (1967). The phenomenon has, in fact, been known to occur in bones since the latter part of the 19th century.

The bodies of people are continuously subjected to varying stresses, but no damage is normally incurred during a day of low level activity. Extended and high levels of activity, such as continuous running, can cause microscopic damage. Live bone has, however, the wonderful capacity to repair or regenerate itself. If the rate of damage accumulation exceeds the rate of repair, however, fatigue cracks can occur (Carter and Hayes, 1976; Wright and Hayes, 1976).

Fatigue tests on compact bone have been conducted by Evans and Lebow (1957), Evans and Riolo (1970) and by Carter and Hayes (1976). Plots of stress versus cycles to failure exhibit a linearly decreasing behavior on log – log diagrams, and tests at different temperatures reveal that fatigue strength is less at 45 °C than at 21 °C.

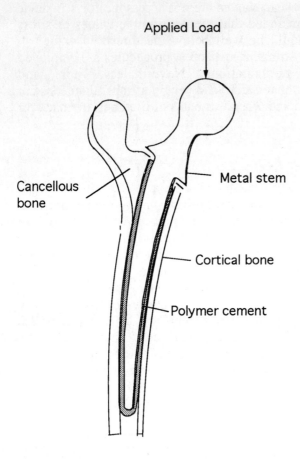

**Fig. 8.1** A sectioned view of an implanted hip joint prosthesis

Carter and Hayes (1977) have also found that fatigue strength decreases with decreasing bone density. This observation is particularly relevant with regard to prosthesis implants in elderly patients. They also observed that progressive fatigue damage resulted in a decrease in stiffness. This behavior differs from that of metals, but it is similar to that which has been found to occur in fiber reinforced composites. It should be noted that the results reported were from tests conducted on dead bone, and the regenerative function of live bone to heal was absent. The results may, therefore, be considered to be conservative.

Wright and Hayes (1976) have applied fracture mechanics procedures to fatigue crack growth and observed Paris law exponents which varied between 3 and 5. The cracks which were monitored propagated parallel to the long axis of the bones and extended along interfaces in the fibrous bone structure. The anisotropic properties of the bone have made it difficult to propagate stable crack growth normal to the long axis. It may be noted that this type of structure resembles that which is encountered in unidirectionally, fiber reinforced composites. A comprehensive review of progress in the characterization of fracture toughness in cortical bone has been presented by Bonfield and Behiri (1989).

An important component in the hip joint prosthesis is the bone cement. The mechanical properties of the cement and its adherence to the bone and to the implant establish the manner in which stress is transferred between the stem and the bone. Obviously, the structural integrity of the cement is crucial to the proper functioning of the remodeled joint.

Polymethyl methacrylate based acrylic cements are widely used in orthopedic surgery. The cements cure in place, but the hand mixing used can entrap bubbles and body fluids and result in some degradation in strength. Some mixtures include additives such as $BaSO_4$ to render them opaque to radiography. Also, antibiotics are sometimes added. The additives may be expected to affect the mechanical properties. Freitag and Cannon (1977) have, however, conducted stress versus cycles to failure tests on bone cements and found that additions of up to 10% of $BaSO_4$ do not significantly affect fatigue strength.

After implantation, there can be a number of voids in the cement and flaws in the cement/bone and cement/metal interfaces. Wright and Robinson (1982) have, therefore, conducted fatigue crack propagation tests to evaluate the damage tolerance of bone cements. Rates of crack growth for stress intensity factors ranging from about 0.5 to 0.33 MPa$\sqrt{m}$

varied from about $10^{-5}$ to $10^{-2}$ mm per cycle. They also conducted crack growth tests on bone cement which was reinforced by short carbon fibers. The resulting composite material had a significantly improved crack growth resistance. When the data were plotted in terms of the ratio of the stress intensity factor range to the effective elastic modulus, however, they found that the data fell on a single line. They suggested that this indicated that the crack growth process was strain controlled.

The relative properties of the three implant components can be expected to have an effect on the manner in which stresses are transferred from the metal stem to the surrounding bone. There is, therefore, a need for research directed toward characterizing the fatigue behavior of the cement/bone and cement/metal interfaces. Experience with interfacial behavior in composites has clearly indicated the importance of this property.

The structural integrity of the hip joint prosthesis should be evaluated by analyses and tests on the total joint. Taylor (1990) has summarized the features of the total problem and has presented the results of a finite element analysis. The joint loads which are encountered depend on the level of activity involved. For standing or slow walking the force is about three times the body weight. For fast walking this factor can rise to as high as seven. In addition to the components described, muscle forces are also a factor.

For early joint prostheses, wear was a problem on the polymeric coating surfaces exposed to bearing forces. Recently, use of high density polyethylene bearing surfaces has improved wear resistance, but the formation of small wear particles may be a cause for concern. Fatigue, however, now appears to be the primary source of failure. As noted previously, cement interfaces affect stress diffusion, and decohesion may adversely affect the distribution of stresses in the components.

A variety of alloys have been used for the stem component. These include stainless steels, cobalt alloys and the titanium 6Al-4V alloy. The fatigue properties of the metals used can be readily obtained. To be useful, however, the stress distribution in the loaded stem must also be known. In calculating stresses it has been found that they are nearly proportional to the elastic modulus (Taylor, 1990). This repeats the behavior cited previously for bone cement, and it indicates that the bone/cement/metal combination is strain controlled. Factors of safety based on the fatigue limit were estimated to range from 3.0 for cobalt alloy to 9.2 for the titanium alloy.

Insight into possible modes of failure can, of course, be extracted from failures of prostheses which have been implanted in patients. Also, interface decohesion is detectable by the use of X-rays. This type of failure does not, however, necessarily result in a failure of the prosthesis. Fatigue failures in the stem can, however, expose a patient to a surgical emergency. Rimnac *et al.* (1986) have examined the failure of a metal hip prosthesis and found beach marks and striations typical of fatigue crack growth. Using curved beam theory and a finite element analysis, they were able to deduce the sequence of loading events which led to failure. The results which they reported led to both a change in subsequent prosthesis design and a change from the use of 316L stainless steel to a cobalt alloy.

Although analyses and tests can be useful in the selection of materials and in design modifications, the loading conditions developed in prostheses are so complex that the degree of uncertainty gravitates against total dependence on them. Also, one of the primary ingredients of the application of damage tolerance concepts is the introduction of scheduled inspections. Since this is either difficult or not possible in implanted prostheses, improvements must depend upon guidance from research and actual implant experience.

# Appendix  A

*R. R. Carlson*

PRODUCT LIABILITY

Modern technology frequently poses grave hazards as well as benefits. Each new generation of machinery presents new possibilities and magnitudes of accident and disaster. Injured workers, consumers and bystanders may seek compensation from the parties responsible for an errant machine or device.

The law of 'product liability' as it exists in most of the developed world evolved in tandem with the industrial revolution and the rise of the consumer society. During the early years of the machine age, judges were often anxious not to discourage nascent industrial enterprises, and were prone to view industrial accidents as acts of nature for which no person was to blame. The prevailing rule was *caveat emptor*. The inventor, manufacturer and seller of any product were safe from liability, unless in selling the product they had made a 'warranty' or express guarantee against the type of defect that happened to cause the accident (Keeton, 1984, section 95A). For a breach of warranty, the buyer could hold the seller liable for the resulting damages.

The early law of warranty was greatly limited by two rules. First, it required proof that the seller did make a warranty, but many sales occurred (as they frequently do today) without a written contract or express warranty. Second, the law required the injured plaintiff to prove he was in 'privity' with the defendant (e.g. the plaintiff was a buyer or party to the sale agreement and was not a mere bystander). An injured person other than the buyer might not be able to recover compensation from the seller, because the seller's guarantee was made to the buyer and no one else (Keeton, 1984, section 96).

As modern machines and consumer goods became more widely dispersed in workplaces, highways and households, the courts became increasingly concerned with the need to protect the public, and they altered the warranty theory of product liability in several ways. First, the courts began to hold that it is **implicit** in any sale transaction that the product in question is fit and safe for its intended use. Thus, an 'implied' warranty is not part of any sale transaction, unless the merchant takes the precaution of expressly disclaiming the warranty (Keeton,1984, section 95A; Howells, 1993).   Second, courts and legislatures began to expand the concept of privity both horizontally and vertically.  The horizontal expansion encompasses certain persons associated with the buyer, such as members of his household.  The vertical expansion encompasses members of the distribution chain such as wholesalers or retailers who have purchased the product and then resold it.

Standing alone, these improvements in the warranty theory would have failed to benefit many persons injured by defective products.  For example, a cautious seller might have disclaimed any warranty (perhaps in the nonnegotiable fine print of a standard form contract), or the plaintiff might be a bystander outside the scope of vertical or horizontal privity with the defendant.  Given these limitations in the warranty theory, the courts developed an alternative 'tort' approach.

In the early tort-based products liability cases, the courts held that a defendant maker or designer of a product could be liable to any foreseeably injured party if the injured party could identify the maker's negligent act or omission that caused the accident.  However, even this theory left some courts dissatisfied, because it is often impossible to prove exactly how a defect occurred or whether it resulted from the maker's failure to use reasonable care.  One answer to this problem has been the development of 'strict liability' theory.

The concept of strict products liability evolved first in the United States over the course of the 20th century (Keeton, 1984, section 98). In contrast with the legal system in the United States, the law of the United Kingdom continued to limit product liability to the negligence and warranty theories described above until 1987.  At that time, however, the United Kingdom also embraced the strict liability theory in the Consumer Protection Act (Devlin, 1990).

Rules of strict liability, which take many different forms, allow an injured party to recover compensation from the maker upon proof that there was a defect that caused the accident.  It is not necessary to prove any particular act or omission that caused the defect.  In this regard, strict liability resembles a warranty, but it constitutes a guarantee to the whole world, and not just to the persons who dealt directly with the maker.

The effect of strict liability can be particularly uncertain in its application to 'design' defects.  Many courts once took the position that a maker/seller could not be liable for any design defect, at least if the alleged shortcoming was 'obvious', or if the maker/seller had sufficiently warned the buyer about the product's hazards. Over time, however, some US courts began to hold that a maker may be liable if the hazards of a particular design outweigh its benefits, or if safer, reasonable alternative designs were available to the maker. Most of these courts have recognized a defense that the 'state of the art' at the time of the product's design and making could not have alerted the maker to the magnitude of the design's hazards or the availability of safer designs (Keeton, 1984, section 99).

Any individual, employee or business involved in the design, making or sale of a defective product could be liable under one or more of the various theories discussed above. This possibility follows from the rule that an agent (including an employee) is not relieved from liability simply because his act or omission was in the course of his services for another person or business (Restatement of Agency, 1957, sections 343, 344 and 350). But could an employee be *strictly* liable (without proof of any personal negligence on his or her part) for his or her  involvement in the design, manufacture or sale of a product? For practical reasons, it would be difficult and extremely unusual for a plaintiff to sue for the strict liability of an individual employee of a corporate producer when the corporate producer has sufficient funds to pay the judgment. However, if the individual employee were held responsible on the basis of 'strict' liability, he or she would be entitled to indemnification from his or her employer if he or she had not been personally negligent with respect to the product in question (Restatement of Agency, 1957, section 439).

Different versions of strict liability are now accepted in most developed nations (Howells, 1993). In western Europe, for example, the European Community has promoted strict liability as the basis for a new minimum standard of consumer protection. In a 1985 directive (Council Directive of July 1985) the Community adopted a strict liability standard for manufacturers of consumer goods, and it directed Member States to enact domestic legislation to implement the standard. Most but not all Member States have now complied with the Directive (Hurd and Zollers, 1993).

The European standard permits considerable latitude for local variation within a general strict liability standard. For example, Member States may choose whether to permit a 'development risks defense', which is the equivalent of the 'state of the art' defense in the United States. Community Members may also impose ceilings on the amount of damages recoverable, provided the ceilings satisfy certain minimum damages standards of the Directive.

In Canada, where product liability laws are established mainly at the provincial level, most courts continue to favor negligence theory, although an increasing number have accepted strict liability theory. Japan continues to limit product liability to breach of warranty theories or tort theories premised on negligence (Rose, 1989).

# Appendix B

SELECTED STRESS INTENSITY FACTOR FORMULAE

Stress intensity factors for the four cases shown in Fig. B.1 can be expressed in terms of the crack lengths, the loading conditions and the geometries of the cracked bodies. References to more extensive collections are made in section 2.4.3. Procedures for determining stress intensity factors for both two and three dimensional cracks are presented in the references cited. The four cases considered have features which are encountered in laboratory testing. They can also serve as a basis for exercises which provide an introduction to the computational procedures which are used in fracture mechanics and fatigue crack growth analyses.

A center cracked plate is shown in Fig. B.1 (a). The stress intensity factor can be expressed as (Fedderson, 1967)

$$K_I = \sigma_\infty \sqrt{\pi a} \left[ \sec(\pi a / 2w) \right]. \tag{B.1}$$

A double edge cracked plate is shown in Fig. B.1 (b). The stress intensity factor for this case can be obtained by use of equation (B.2) (Keer and Freedman, 1973).

$$K_I = \sigma_\infty \sqrt{\pi a} \left[ 1.12 - 0.61(a/w) + 0.13(a/w)^3 \right] \left[ 1 - (a/w) \right]^{-0.5}. \tag{B.2}$$

The stress intensity factor for the plate with a single edge crack in Fig. B.1 (c) can be expressed as (Gross, Srawley and Brown, 1964)

$$K_I = \sigma_\infty \sqrt{\pi a} \left[ 1.12 - 0.23(a/w) + 10.6(a/w)^2 - 21.7(a/w)^3 + 30.4(a/w)^4 \right]. \tag{B.3}$$

The stress intensity factor for the cracked bend specimen shown in Fig. B.1 (d) can be determined by use of equation (B.4) (Gross and Srawley, 1972).

$$K_I = CPW \left[ B(w-a) \right]^{-1.5}. \tag{B.4}$$

The coefficient $C$ varies from 3.69 for $a/W = 0.4$ to 3.81 for $a/W = 0.6$. $B$ in equation (B.4) is the beam thickness.

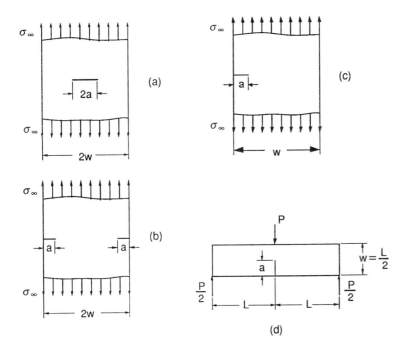

**Fig. B.1** (a) A center cracked specimen, (b) a double edge cracked specimen, (c) a single edge cracked specimen and (d) a single edge cracked bend specimen

# Problems

## Chapter 1

1.1 In the cyclic range $10^3 < N < 10^6$ a log – log plot of stress versus cycles to failure can be approximated by a straight line. A power relationship of the form $\sigma^m N = C$ can be used, where $m$ and $C$ are constants. Given the data values of $\sigma_1 = 630$ MPa for $N_1 = 10^3$ and $\sigma_2 = 350$ MPa for $N_2 = 10^6$ determine the values of $m$ and $C$. Discuss what might be rational choices for design stresses for the ranges of $N < 10^3$ and $N > 10^6$.

1.2 Rewrite the equation of problem 1.1 in the form $(\sigma/\sigma_e)^n (N/10^6) = k$ where $\sigma_e$ is the endurance limit. Determine $n$ and $k$. Use the $\sigma$ and $N$ values of problem 1.1 and set $\sigma_e = 300$ MPa.

1.3 Describe a rational procedure for choosing the values given in problem 1.1. Suppose that for the material of interest, there is an enduance limit which is known, and that you also know the value of the ultimate strength. What factors may influence the selection of a factor of safety?

## Chapter 2

2.1 Use equation (2.5) to obtain a relation between the strain $\varepsilon$ and the nominal strain $e$ which is defined as $e = (L - L_0)/L_0$.

2.2 Obtain a relationship between the strain $\varepsilon$ and the cross-sectional areas $A$ and $A_0$. Assume that there is no change in volume in going from the unloaded to the loaded state, i.e. $AL = A_0 L_0$.

2.3 Using equation (2.5), make a plot of $L/L_0$ versus $\varepsilon$. Include both tensile and compressive values of $\varepsilon$.

2.4 Use the result from problem 2.2 to plot $A/A_0$ versus $\varepsilon$ for both tension and compression.

2.5  A necessary condition for the attainment of a maximum tensile force is $dF = 0$. Use the results of the preceding problems to show that  the derivative $d\sigma/d\varepsilon = \sigma$. Note that $F = \sigma A$.

2.6  A plot of log $\sigma$ versus log $\varepsilon$ is often a straight line. This suggests the use of a stress – strain relationship of the form $\sigma = B\varepsilon^n$.  Show that for this type of behavior the strain at the maximum load $\varepsilon_{max} = n$.

2.7  Derive equilibrium equations (2.14) and (2.15).

2.8  Show that the shear stresses $\sigma_{12}$ and $\sigma_{21}$ are equal. (Assume no couple stresses are present).

2.9  Consider a two dimensional problem in which couple stresses are developed by distributed moments per unit area, $m_1$ and $m_2$, on the edges, $\Delta x_1$ and $\Delta x_2$, of a free body element. How is the equation $\sigma_{12} = \sigma_{21}$ modified by the presence of the couple stresses?

2.10  Use the relations for $\varepsilon_{11}$, $\varepsilon_{22}$ and $\varepsilon_{12}$ in equations (2.21) and (2.22) to eliminate $u_1$ and $u_2$ and obtain equation (2.24).

2.11  Invert equations (2.25) to obtain equations for the stress components in terms of the strain components.

2.12  Consider the case of plane stress in which $\sigma_{33} = \sigma_{13} = \sigma_{23} = 0$ and modify equations (2.25) for an orthotropic plate.

2.13  Derive Hooke's laws for homogeneous, isotropic solids for plane stress (stress components with a subscript of 3 are zero) and plane strain (strain components with a subscript of 3 are zero).

2.14  Show that the Airy stress function satisfies the equilibrium equations for two dimensional problems.

2.15  Use a two dimensional free body element to solve for the stress components in polar coordinates in terms of those in Cartesian coordinates.

**2.16** Write the equations for the stress components in equation (2.34) for $n = 3$.

**2.17** Perform the indicated integration in equation (2.45) to obtain equation (2.46).

## Chapter 3

**3.1** Two tensile specimens of a brittle material have a volume ratio, $V_2/V_1$, of 4. Use equation (3.1) to determine the ratio of the ultimate strengths for a value of $m = 4$.

**3.2** Use equation (3.2) to determine the ratio of the tensile and bending strengths for specimens of the same size. Use $m = 4$.

**3.3** The rate of crack growth in the near threshold region increases rapidly with increasing cyclic plastic zone size. Discuss how this might effect analytical predictions of crack growth rate in this region.

**3.4** A material with a time dependent response behaves as though it were a linear spring and dashpot in series. The spring response can be written as $\sigma = E\varepsilon$ where $\varepsilon$ and $\sigma$ are the strain and stress, respectively, and $E$ is an elastic constant. The dashpot response is $\lambda\, d\varepsilon/dt = \sigma$ where $d\varepsilon/dt$ is the rate of strain in the dashpot and $\lambda$ is a constant. Show that the constitutive law can be written as $E\lambda\, d\varepsilon/dt = E\sigma + \lambda\, d\sigma/dt$.

**3.5** Show that the constitutive law for a material which behaves as if it were a linear spring and a dashpot in parallel is $\sigma = \lambda\, d\varepsilon/dt + E\varepsilon$.

**3.6** A creep test is conducted by applying a constant stress $\sigma_0$ for $t \geq 0$. Show that the strain history is described by $\varepsilon = (\lambda + Et)\sigma_0/\lambda E$ for the viscoelastic law of problem 3.4.

**3.7** Use the constitutive law of problem 3.4 to determine the strain history for a test in which $\sigma = \sigma_m + \sigma_a \sin \omega t$.

**3.8** A single continuous fiber with an elastic modulus $E_f$ is embedded in the center of a cylinder with an elastic modulus $E_m$. When a uniaxial load is applied, the interface between the components remains intact, so it can be assumed that the strains are equal. Show that an effective elastic modulus, $E_e$, of the composite is $E_e = (E_f A_f + E_m A_m)/A$ where $A_f$ and $A_m$ are the fiber and matrix cross-sectional areas and $A = A_f + A_m$.

3.9 Discuss how the effective stiffness of a unidirectional lamina would vary with orientation for inplane loading. For what orientation would stiffness be smallest?

## Chapter 4

4.1 The equation in problem 1.1 is to be used in the design of an engine component. Derive an equation for estimating the error $\Delta N$ in the number of cycles to failure if the estimated stress is in error by an amount $\Delta \sigma$. Determine the percent error in $N$ if $\Delta \sigma = \pm 0.02\sigma$ and $m = 10$.

4.2 For low cycle fatigue a relationship between the alternating strain $\varepsilon_a$ and the cycles to failure $N$ has the form $\varepsilon_a = 0.5\varepsilon_f(N)^{-\frac{1}{2}}$ where $\varepsilon_f$ is the logarithmic strain at fracture for monotonic loading in tension. What type of plot would be effective for varifying the validity of this equation? If the estimated value of $\varepsilon_a$ in an application were 10 percent low, what would be the ratio of the actual to the predicted number of cycles to failure? Would this error be considered acceptable if the predicted value of $N$ was 800 cycles? How might the possibility of such errors affect design?

4.3 Equation (4.8) provides a method for determining the effect of mean stress. How might the diagram which illustrates this equation be modified to take into account values of $\sigma_m < 0$? If data for $\sigma_m < 0$ are available, how might an equation for negative $\sigma_m$ be developed?

4.4 If a long slender rod is subjected to cyclic loading with $\sigma_m < 0$, and buckling is not acceptable, how might the diagram from problem 4.3 be modified to exclude buckling? What kind of a curve could be used as a design limit? Suppose that $\sigma_e > |\sigma_c|$. How might this affect the diagram?

**4.5** If the loading in a notched component does not produce stresses which exceed the elastic limit, a safe life design approach can be used. The use of equations (4.3) and (4.4), however, requires a knowledge of the value of $q$. Peterson (1959) has suggested an empirical equation for $q$ of the form $q = (1 + a/r)^{-1}$ for $0 \leq q \geq 1$ where $r$ is the notch radius and $a$ is a material constant which must be obtained from fatigue data on notched and unnotched tests. Substitution of this empirical relation into equation (4.4) gives $K_f = 1 + (K_t - 1)(1 + a/r)^{-1}$. Given a $K_t = 4$ and an $r = 1.0$ mm, show that $K_f$ is relatively insensitive to variations in $a$ for a hardened steel for $a$ values of the order 0.025 mm.

**4.6** For a closed, thin-walled, cylindrical pressure vessel the stress in the circumferential direction is twice that for the longitudinal direction. Use equation (4.10) to determine the internal pressure at which yielding begins. First, derive the equations for the longitudinal and circumferential stresses in terms of the pressure $p$, the wall thickness $t$, and the cylinder radius $R$. Assume that the radical stress is negligible for $t/R \ll 1$.

**4.7** A bar with a circular cross-section of radius $r$ and length $L$ is fixed at one end, and loaded at the other end by a cyclic force $P$ applied to a cross-bar of length $e$. The load ratio $R = -1$. The force applies a moment $M = PL$ at the fixed end and a torque $T = Pe$ along the bar. Determine the location at which the maximum stresses are developed, and use equation (4.13) to derive an equation for $P$ in terms of $\sigma_e$ and the geometrical properties of the bar. Assume that the ratio $L \gg r$ so that direct shear stresses can be neglected.

**4.8** The load spectrum for the cyclic loading of an aluminium alloy component has been reduced to a repeating block loading with three stress levels. The stress levels, and the number of cycles to failure are given in the tabulation below.

| Stress (MPa) | Failure (cycles) | Cycles per block |
|---|---|---|
| 300 | $7 \times 10^4$ | $1 \times 10^3$ |
| 210 | $2 \times 10^6$ | $8 \times 10^3$ |
| 250 | $7 \times 10^5$ | $4 \times 10^3$ |

Use the Palmgren – Miner damage rule to determine the number of cycles to failure. How many blocks would be completed before failure?

4.9 Identify some of the limitations of the Palmgren – Miner rule and describe service applications in which it may be reasonably acceptable.

4.10 Describe the phenomena of cyclic strain hardening and softening.

4.11 Outline a procedure for obtaining stabilized cyclic stress – strain diagrams.

4.12 A heat treated steel is to be exposed to cyclic straining in service. The constants in equations (4.21), (4.22) and (4.23) are as follows: $E = 200$ GPa, $\sigma'_f = 2200$ MPa, $\varepsilon'_f = 0.11$, $b = -0.08$, $c = -0.65$. Plot $0.5\Delta\varepsilon_E$, $0.5\Delta\varepsilon_p$ and $0.5\Delta\varepsilon_T$ versus $2N_f$ on log – log coordinates.

4.13 In equation (4.24) the stress and strain concentration factors are defined as $K_\sigma = \sigma/S$ and $K_\varepsilon = \varepsilon/e$ where $S$ and $e$ are the nominal stresses and strains and $\sigma$ and $\varepsilon$ are the actual stresses and strains. Locate the coordinates $(S, e)$ and $(\sigma, \varepsilon)$ on a stress – strain diagram for a linear elastic material. Locate these coordinates on the diagram for an elastic – plastic material for which $(S, e)$ is within the elastic range and $(\sigma, \varepsilon)$ have been subject to yielding. Note that for the latter case, $K_\sigma$ is smaller and $K_\varepsilon$ is larger than they would be for elastic loading. As defined in equation (4.24), $K_t$ is the geometric mean of $K_\sigma$ and $K_\varepsilon$.

4.14 How must the function exp $(-0.5x^2)$ be modified for use as a probability density function? (a) Multiply this function by $(2\pi)^{-\frac{1}{2}}$ and determine the first area moment about $x = 0$ to show that the mean value of $x$ is zero. (b) Determine the second area moment about $x = 0$ to show that the variance or standard deviation is unity. Integrate from $-\infty$ to $+\infty$.

4.15 Plot the probability density function given in problem 4.14. If you differentiate this function, what do you obtain? (a) What would be the probability of occurrence for $-0.5 < x < +0.5$? (b) What would be the probability of occurrence for $-0.1 < x < +1.0$? (c) What values of $a$ for $-a < x < +a$ would give a probability of occurrence of 0.95?

Chapter 5

5.1 The stress intensity factor for a tensioned plate with an edge crack $a$ which is very much smaller than the plate width can be written as $K_I = \sqrt{\pi a}$. Substitute this into equation (5.3) for cyclic loading with $R = 0$ and determine the relationships between the exponents of equations (5.2) and (5.3).

5.2 Show that the denominator of equation (5.4) can be written in the form shown in equation (5.5).

5.3 Show that equation 5.6 is equivalent to equation (5.4).

5.4 Is the rate of crack growth in a fatigue test a continuous function of $N$? Write equation (5.3) as a finite difference equation.

5.5 Show that equation (5.10) can be obtained by integration of equation (5.9) For this problem $n = 3$.

5.6 Cite an example for which equation (5.12) does not provide an accurate estimate of the plastic zone size. Consider both long and small cracks.

5.7 Show that as the ratio of $\Delta K_{th}$ to $\Delta K$ becomes small, the crack growth rate predicted by equation (5.15) approaches that for equation (5.3).

5.8 Are the 'intrinsic' and 'extrinsic' threshold ranges of stress intensity factors independent of one another or coupled? Explain your answer.

5.9 According to the discrete asperity model, closure behavior is represented as a straight line on a plot of stress intensity versus external load (Figure 5.3). Obtain end points of the line by setting the external load equal to zero (point D) and the internal, closure load equal to zero (point B). Show that point D is given by equation (5.23) and point B is given by equation (5.24).

5.10 The 'delay distance' for the Willenborg overload model can be determined by setting $K_R$ in equation (5.25) equal to zero. Obtain an equation for the ratio of $a$ to $\rho_{OL}$ for $n = 1,2,3$ for the equation below.

$\left[ (K_{OL})^n (1 - a/\rho_{OL})^n - (K_{max})^n \right]^{1/n} = 0$. Use a ratio of $K_{max}$ to $K_{OL}$ of 0.5.

5.11 Indicate by a diagram how the ratio of $\Delta K_{eff}$ to $\Delta K$ varies with the ratio of $S_o$ to $S_{max}$ in equation (5.30). Take $R = 0$. How would positive and negative values of $R$ affect the plot? What is a limitation on the use of negative $R$?

5.12 Would the use of equation (5.33) for 'long' cracks give results which would be compatible with those of equation (5.3)? Consider the case of $R = 0$, and let $K \propto \sigma \sqrt{a}$.

5.13 Discuss the application of the mixed mode fatigue crack rule proposed by Pook (equation 5.47). What type of behavior is predicted on either side of the curve?

5.14 Discuss how coupling effects between fatigue and creep and between fatigue and corrosion might affect crack growth.

Chapter 6

6.1 Describe how the probability density functions for crack size and for load at a given location affect the risk failure. How are inspection intervals affected by these factors? For what conditions can a crack which is too small to be detected become critical?

6.2 Make a list of types of sites which could possibly be represented by an 'equivalent initial flaw size'.

6.3 Given a probability density function of the form $f = x$ on $0 \le x \le \sqrt{2}$, determine the cumulative distribution function of $x$. Plot both functions.

6.4 Perform the same analyses and plots of problem 6.3 for the probability density function $f = 1.817x - x^2$ on $0 \le x \le 1.817$.

6.5 Compare the features of the crack growth law of equation (6.2) with those of a Paris type law.

6.6 Use equation (6.2) with $b = 1$ and equation (6.1) to obtain equation (6.3). Let $a(o) = x(o)$ and $a(t) = x(t)$.

6.7 Discuss the implications of the growth features of the deterministic law of equation (6.2) and the stochastic law of equation (6.4).

## Chapter 7

7.1 Calculate the local strain concentrations in the $\pm 25°$ plies of a $[\pm 25/90_n]_s$ laminate for $n = 4, 6, 8$ for (a) an edge delamination only, (b) a local delamination only and (c) a combined edge and local delamination. The delaminations occur at the interfaces between the $90°$ plies and adjacent angle plies. Use lamina properties for the $0°$ ply as follows: $E_{11} = 134$ GPa, $E_{22} = 10.2$ GPa, $G_{12} = 5.52$ GPa, $v_{12} = 0.30$. The plies have the same thickness.

7.2 Calculate the local strain concentrations in the $\pm \theta°$ plies of a $[\pm \theta/90_n]_s$ laminate for $\theta = 35°, 45°$ and $n = 4, 6, 8$ and for (a) edge delamination only, (b) local delamination only and (c) combined edge and local delamination. Use lamina properties as in problem 7.1. The delaminations occur again at the interfaces between the $90°$ plies and adjacent angle plies.

7.3 Calculate the stiffness loss due to edge delamination at the $-25/90$ interfaces in a $[\pm 25/90_2]_s$ laminate. The delamination length to laminate halfwidth ratio $a/b$ is 0.30. Use lamina properties as in problem 7.1.

7.4 Calculate the stiffness loss due to local delamination (delaminations growing from matrix cracks) at the $-25/90$ interfaces in a $[\pm 25/90_2]_s$ laminate. The delamination length to laminate length ratio, $a/l$, is 0.30. Use lamina properties as in problem 7.1.

7.5 Using equation (7.9) for the energy release rate, derive an equation for the number of cycles for the first delamination, $N_1$, to form when the laminate is cycled with peak stress $\sigma_{max}$. The critical energy release rate is $G_c$.

7.6 Calculate the number of cycles for the initiation of each successive local delamination $N_i$, i.e. derive equation (7.11).

# References

Adam, T., Gathercole, N., Harris, B. *et al.* (1991) The fatigue behavior of a CFRP laminate subjected to load cycling at *R* ratios from pure tension to pure compression, in *Proceedings 8th International Conference on Composite Materials* (eds S.W. Tsai and G.S. Springer), Society for the Advancement of Materials and Process Engineering, Covina, Ca, **4** (38), 1 – 7.

Adams, D.F. and Walrath, D.E. (1987) Further development of the Iosipescu Shear Test Method, *Experimental Mechanics*, **27**, 113 – 19.

Agarwal, B.D. and Broutman, L.J. (1990) *Analysis and Performance of Fiber Composites*, John Wiley, New York.

Alden, T.H. and Backofen, W.A. (1961) The formation of fatigue cracks in aluminum single crystals, *Acta Metalurgica*, **9**, 352 – 66.

Almen, J.O. and Black, P.H. (1963) *Residual Stresses and Fatigue in Metals*, McGraw-Hill, New York.

Amateau, M.F., Conway, J.C. Jr. and Bhagat, R.B. (1993) Ceramic composites: design, manufacture and performance, in *Proceedings 9th International Conference on Composite Materials*, Vol. 2 (ed. A. Miravete), Woodhead Publishing, Cambridge, UK, pp. 23 – 31.

Ashby, M.F. (1972) A first report on deformation mechanism maps, *Acta Metalurgica*, **20**, 887 – 97.

Ashton, J.E., Halpin, J.C. and Petit, P. (1969) *Primer on Composite Materials: Analysis*, Technomic, Westport, Ct.

ASTM Standard Designation E 399-74 (1974) *Standard Method of Test for Plane-Strain Fracture Toughness of Metallic Materials*, ASTM Standards, American Society for Testing and Materials.

ASTM Test Method E647 (1993) *Proposed Small Cracks Appendix, Version 3.000*, American Society for Testing and Materials, pp. 1 – 8.

Atluri, S.N., Sampath, S.G. and Tong, P. (1990) *Structural Integrity of Aging Airplanes*, Springer-Verlag, Berlin.

Atluri, S.N., Sampath, S.G. and Tong, P. (1991) Computational schemes for integrity analyses of fuselage panels in aging airplanes, in *Structural Integrity of Aging Airplanes*, (eds S.N. Atluri, S.G. Sampath and P. Tong) Springer-Verlag, Berlin, pp. 15 – 35.

Atluri, S.N., Harris, C.E., Hoggard, A., *et al.* (1992) *Durability of Metal Aircraft Structures*, Atlanta Technical Publishers, Atlanta, Ga.

Badaliance, R. (1981) Mixed mode fatigue crack propagation, in *Mixed Mode Crack Propagation* (eds G.C. Sih and P.S. Theocaris), Sijthoff and Noordhoff, The Netherlands, pp. 77 – 98.

Baden-Powell, D.F.W. (1949) Experimental clactonian technique, *Proceedings of the Prehistoric Society*, **15**, 38 – 41.

Baker, A.A. and Jones, R. (eds) (1988) *Bonded Repair of Aircraft Structures*, Martinus Nijhoff, Dordrecht, The Netherlands.

Baloch, R.A. and Brown, M.W. (1993) Crack closure analysis for the threshold of fatigue crack growth under mixed-mode I/II loading, in *Mixed-Mode Fatigue and Fracture* (eds H.P. Rossmanith and K.J. Miller), Mechanical Engineering Publications, London, pp. 125 – 37.

Bannantine, J.A., Comer, J.J. and Handrock, J.L. (1990) *Fundamentals of Metal Fatigue*, Prentice-Hall, Englewood Cliffs, NJ.

Bao, G., Fan, B. and Evans, A.G. (1992) Mixed mode delamination cracking in brittle matrix composites, *Mechanics of Materials*, **13**, 59 – 66.

Barney, C., Cardona, D.C. and Bowen, P. (1993) Predictions of crack arrest under cyclic loading in continuous fibre reinforced metal matrix composites, in *Proceedings 9th International Conference on Composite Materials* (ed. A. Miravete), Woodhead Publishing, Cambridge, UK, pp. 541 – 8.

Basquin, O.H. (1910) The exponential law of endurance tests, *Proceedings of the American Society for Testing and Materials*, **10**, 625 – 30.

Bassini, J.L. and McClintock, F.A. (1981) Creep relaxation of stress around a crack tip, *International Journal of Solids and Structures*, **17**, 479 – 92.

Bates, R.C. and Clark, W.G. (1969) Fractography and fracture mechanics, *American Society for Metals Transactions Quarterly*, **62**, 380 – 9.

Bathias, C. and Pelloux, R.M. (1973) Fatigue crack propagation in martensitic and austenitic steel, *Metallurical Transactions*, **4**, 1265.

Bauschinger, J. (1886) *On the Change of the Position of the Elastic Limit of Iron and Steel Under Cyclic Variations of Stress*, Mitt. Mech. Tech. Lab., Munich, **13**, 1.

Beevers, C.J. (ed.) (1980) *The Measurement of Crack Length and Shape During Fracture and Fatigue*, EMAS Ltd, Warley, UK.

Beevers, J.C. and Carlson, R.L. (1986) A consideration of significant factors controlling fatigue thresholds, in *Fatigue Crack Growth – 30 Years of Progress* (ed. R.A. Smith), Pergamon Press, Oxford, pp. 89 – 101.

Beevers, C.J., Cooke, R.J., Knott, J.F. *et al.* (1975) Some considerations of the influence of sub-critical cleavage growth during fatigue crack propagation in steels, *Metal Science,* **9**, 119 – 26.

Beevers, C.J., Bell, K., Carlson, R.L. *et al.* (1984) A model for fatigue crack closure, *Engineering Fracture Mechanics*, **19**, 93 – 100.

Béland, S. (1990) *High Performance Thermoplastic Resins and Their Composites*, Noyes Data Corp., Park Ridge, N J.

Berens, A.P., Hovey, P.W. and Skinn, D.A. (1991) *Risk Analysis for Aging Aircraft Fleets*, Vol. 1 – Analysis, UDRI-TR-91-43, University of Dayton Research Institute, Dayton, Oh.

Beuth, J.L. and Hutchinson, J.W. (1994) Fracture analysis of multi-site cracking in fuselage lap joints,*Computational Mechanics*, **13**, pp. 315 – 31.

Blom, A.F. (1993a) Fatigue crack growth modeling: procedures and problems, in *Theoretical Concepts and Numerical Analysis of Fatigue* (eds. A.F. Blom and C.J. Beevers), EMAS, Warley, UK, pp. 439 – 61.

Blom, A.F. (ed.) (1993b) *Durability and Structural Integrity of Airframes*, EMAS, Warley, UK.

Blom, A.F., Hedlund, A., Zhao, W. *et al.* (1986) A short fatigue growth behavior in Al 2024 and Al 7475, in *The Behavior of Short Fatigue Cracks* (eds K.J. Miller and E.R. de los Rios), Mechanical Engineering Publ., London, pp. 37 – 66.

Bokros, J.C. (1977) Carbon biomedical devices, *Carbon*, **15**, 355 – 71.

Bolingbroke, R.K. and King, J.E. (1986) A comparison of long and short fatigue crack growth in a high strength aluminum alloy, in *The Behavior of Short Fatigue Cracks* (eds K.J. Miller and E.R. de los Rios), Mechanical Engineering Publ., London, pp. 101 – 14.

Boller, K.H. (1957) Fatigue properties of fibrous glass reinforced plastic laminates subjected to various conditions, *Modern Plastics*, **34**, 163 – 88.

Bonfield, W. and Behiri, J.C. (1989) Fracture toughness of natural composites with reference to cortical bone, in *Application of Fracture Mechanics to Composite Materials* (ed. K. Friedrich), Elsevier Science Publishers, Amsterdam, pp. 615 – 35.

Bordia, R.J., Dalgleish, B.J., Charalambides, P.G. *et al.* (1991) Cracking and damage in a notched unidirectional fiber reinforced brittle matrix composite, *Journal of the American Ceramic Society*, **4**, 2776 – 80.

Bossman, J. and Lof, C.J. (1985) *Technical Description and Users' Guide of Program PRECRAC*, National Aerospace Laboratory NLR Memorandum, Amsterdam, The Netherlands.

Bøving, K.G. (1989) *NDE Handbook*, Butterworths, London.

Bower, A.F. and Ortiz, M. (1990) Solution of three dimensional crack problems by finite perturbation method, *Journal of the Mechanics and Physics of Solids*, **82**, 45 – 57.

Braithwaite, F. (1854) On the fatigue and consequent fracture of metals, *Proceedings of the Institute of Civil Engineers*, **13**, 463 – 75.

*British Standards Institution B5 5762-1979 (1979) Methods for Crack Opening Displacement (COD) Testing,* British Standards Institution, London.

Broek, D. (1982) *Elementary Engineering Fracture Mechanics*, 3rd edn, Nijhoff, The Hague.

Brown, M.W. and Miller, K.J. (eds) (1989) *Biaxial and Multiaxial Fatigue*, Mechanical Engineering Publ., London.

Bucci, R.J. (1970) *Environment Enhanced Fatigue and Stress Corrosion Cracking of a Titanium Alloy Plus a Simple Model for Assessment of Environmental Influence on Fatigue Behavior*, Ph.D. Thesis, Lehigh Univ., Bethlehem, Pa.

Bucci, R.J. (1981) Development of a proposed ASTM standard test method for near threshold fatigue crack growth rate measurement, in *Fatigue Crack Growth Measurement & Data Analysis* (eds S.J. Hudak, Jr. and R.J. Bucci), ASTM STP 738, American Society for Testing and Materials, Philadelphia, pp. 5 – 28.

Bucci, R.J., Brazill, R.L., Sprowls, D.O *et al.* (1986) The breaking load method: a new approach for assessing resistance to growth of early stage stress corrosion cracks, in *Proceedings International Conference and Exposition on Fatigue, Corrosion Cracking, Fracture Mechanics and Failure Analysis* (ed. V.S. Goel), American Society for Metals, Metals Park, Oh, pp. 267– 77.

Buch, A., Seeger, T. and Vormald, M. (1986) Improvement of fatigue life prediction by use of correction factors, *International Journal of Fatigue*, **8**, 175 – 85.

Buck, O., Thompson, R. and Rehbein, D. (1988) Using Acoustic Waves for the Characterization of Closed Fatigue Cracks, in *Mechanics of Fatigue Crack Closure* (eds J. C. Newman Jr. and W. Elbert), ASTM STP 982, American Society for Testing and Materials, Philadelphia, pp. 536 – 547.

Budiansky, B. and Fleck, N.A. (1993) Compressive failure of fibre composites, *Journal of Mechanics and Physics of Solids*, **41**, 183 – 211.

Budiansky, B. and Hutchinson, J.W. (1978) Analysis of closure in fatigue crack growth, *Journal of Applied Mechanics*, **45**, 267 – 76.

Budiansky, B., Hutchinson, J.W. and Evans, A.G. (1986) Matrix fracture in fiber-reinforced ceramics, *Journal of the Mechanics and Physics of Solids*, **34**, 167 – 89.

Budiansky, B., Hutchinson, J.W. and Lambropoulos, J.C. (1983) Continuum theory of dilatant transformation toughening in ceramics, *International Journal of Solids and Structures*, **19**, 337 – 55.

Campbell, G.H., Ruhle, M., Dalgleish, B.J. *et al.* (1990) Whisker toughening: a comparison between aluminum oxide and silicon nitride toughened with silicon carbide, *Journal of the American Ceramic Society*, **73**, 521 – 30.

Cardon, A.H. and Verchery, G. (eds) (1990) *Durability of Polymer Based Composite Systems for Structural Applications*, Elsevier Applied Science, London.

Cardona, D.C., Bowen, P. and Beevers, C.J. (1992) Fatigue crack growth in zirconia partially stabilized with ceria (Ce-TZP), in *Theoretical Concepts and Numerical Analysis of Fatigue* (eds A.F. Blom and C.J. Beevers), EMAS, Warley, UK, pp. 571 – 89.

Carlson, R.L. (1989) Crack extension in metals with fracture toughness gradients, *International Journal of Fracture*, **41**, R51 – 4.

Carlson, R.L. and Beevers, C.J. (1984) A multiple asperity fatigue crack closure model, *Engineering Fracture Mechanics*, **20**, 687 – 90.

Carlson, R.L. and Beevers, C.J. (1985) A mixed mode fatigue crack closure model, *Engineering Fracture Mechanics*, **22**, 651 – 60.

Carlson, R.L. and Beevers, C.J. (1992) Effects of overloads and mixed modes on closure, in *Theoretical Concepts and Numerical Analysis of Fatigue* (eds. A.F. Blom and C.J. Beevers), EMAS, Warley, UK, pp. 277 – 97.

Carlson, R.L. and Kardomateas, G.A. (1994) Effects of compressive load excursions on fatigue crack growth, *International Journal of Fatigue*, **16**, 141 – 6.

Carlson, R.L., Kardomateas, G.A. & Bates, P.R. (1991) On the effects of overloads in fatigue crack growth, *International Journal of Fatigue*, **13**, 453 – 60.

Carlson, R.L., Blakeley, E., Kardomateas, G.A. *et al.* (1993) Effects of negative *R* values on fatigue crack growth, in *Proceedings 5th International Conference on Fatigue and Fatigue Thresholds* (eds J.P. Bailon and J.I. Dickson), EMAS, Warley, UK, pp. 877 – 82.

Carlsson, L.A. (1990) *Thermoplastic Composite Materials*, Elsevier, London.

Carlsson, L.A. and Gillespie, J.W. Jr. (1989) Mode II interlaminar fracture of composites, in *Application of Fracture Mechanics to Composite Materials* (ed. K. Friedrich), Elsevier Science Publishers, Amsterdam, pp. 113 – 57.

Carlsson, L.A. and Pipes, R.B. (1987) *Experimental Characteristics of Advanced Composites*, Prentice-Hall, Englewood Cliffs, NJ.

Carter, D.R. and Hayes, W.C. (1976) Fatigue life of compact bone - effects of stress amplitude, temperature and density, *Journal of Biomechanics*, **9**, 27– 34.

Carter, D.R. and Hayes, W.C. (1977) Compact bone fatigue damages - residual strength and stiffness, *Journal of Biomechanics*, **20**, 32 – 7.

Caslini, M., Zanotti, C. and O'Brien, T.K. (1987) Study of matrix cracking and delamination in glass/epoxy laminates, *Journal of Composites Technology and Research*, **9**, (4), 121 – 30.

Chai, H. (1992) Experimental evaluation of mixed-mode fracture in adhesive bonds, *Experimental Mechanics*, **32,** pp. 296 – 303.

Chamis, C.C. (1974) Mechanics of load transfer at the interface, in *Composite Materials,*,Vol. 6, (ed. E.P. Plueddeman), Academic Press, New York, pp. 31 – 77.

Chamis, C.C. and Sinclair, J.H. (1977) Ten-degree off-axis test for shear properties in fiber composites, *Experimental Mechanics*, **17**, 339 – 46.

Chan, W.S., and Ochoa (1988) Suppression of edge delamination in composite laminates by terminating a critical ply near the edges, in *Proceedings 29th AIAA/ASME/ASCE/AHS/ASC Structures*, Structural Dynamics and Materials Conference, Washington, D.C., pp. 359 – 64.

Chan, W.S., Rogers, C. and Aker, S. (1986) Improvement of edge delamination strength of composite laminates using adhesive layers, in *Composite Materials Testing and Design* (ed. J.M. Whitney), ASTM STP 893, American Society for Testing and Materials, Philadelphia, pp. 266 – 82.

Chang, J.B. (1981) Round-robin crack growth predictions on center-cracked tension specimens under random spectrum loading, in *Methods and Models for Predicting Fatigue Crack Growth under Random Loading* (eds J.B. Chang and C.M. Hudson), ASTM STP 748, American Society for Testing and Materials, Philadelphia, pp. 3 – 39.

Chang, J.B. and Hudson, C.M.(eds) (1981) *Methods and Models for Predicting Fatigue Crack Growth under Random Loading*, ASTM STP 748, American Society for Testing and Materials, Philadelphia.

Charalambides, P.G. and McMeeking, R.M. (1987) Finite element method simulation of crack propagation in a brittle micro-cracking solid, *Mechanics of Materials*, **6**, 71 – 87.

Chen, A.S., Harris, B. and Reiter, H. (1991) The effect of fatigue damage on the mechanical properties of CFRP composites, in *Proceedings 8th International Conference on Composite Materials*, (eds S.W. Tsai and G.S. Springer), Society for the Advancement Material and Process Engineering, Covina, Ca, pp. 38F 1 – 9.

Christensen, R.H. (1959) *Metal Fatigue*, McGraw-Hill, New York.

Christensen, R.H. (1963) Fatigue crack growth affected by metal fragments wedged between opening – closing crack surfaces, *Applied Materials Reseach*, October, pp. 207 – 10.

Chu, H.P. and Wacker, G.A. (1969) Stress corrosion testing of 7079-T6 aluminum alloy on sea water using smooth and precracked specimens, Trans. ASME, *Journal of Basic Engineering*, **91D**, 565 – 9.

Clarke, G.A., Andrews, W.R., Paris, P.C. *et al.* (1976) Single specimen tests for $J_{IC}$ determination, in *Mechanics of Crack Growth*, ASTM STP 590, American Society for Testing and Materials, Philadelphia, pp. 27 – 42.

Claussen, N. (1981) Design of transformation-toughened ceramics, *Advances in Ceramics, Science and Technology of Zirconia*, Vol. 12, (eds A.H. Heuer and L.W. Hobbs), American Ceramic Society, Columbus, Oh, pp. 137 – 63.

Coffin, L.F. (1954) A study of the effects of cyclic thermal stresses on a ductile metal, *Transactions American Society of Mechanical Engineers*, **76**, 931 – 50.

Collins, J.A. (1981) *Failure of Materials in Mechanical Design*, Wiley-Interscience, New York.

Council Directive of July 1985 on the approximation of the laws, regulations and administrative provisions of the Member States concerning liability for defective products, 28 O.J. Eur. Comm. (No. L210) 29.

Cox, B.N. and Marshall, D.B. (1990) Analogies between bridged cracks in fatigue and monotonic loading, in *Proceedings 4th International Conference on Fatigue and Fatigue Thresholds* (eds H. Kitagawa and T. Tanaka), MCEP, Birmingham, UK, pp. 659 – 64.

Crossman, F.W. and Wang, A.S.D. (1992) The dependence of transverse cracking and delamination on ply thickness in graphite/epoxy laminates, in *Damage in Composite Materials* (ed. K.L. Reifsnider), ASTM STP 775, American Society for Testing and Materials, Philadelphia, pp. 118 – 39.

Cruse, T.A. (1975) *Boundary Integral Equation Method for Three Dimensional Elastic Fracture Mechanics Analysis*, AFOSR-TR-75-0813, Accession No. A0A 011660.

Dauskart, R.H., Marshall, D.B. and Ritchie, R.O. (1990) Cyclic fatigue crack propagation in magnesia – partially – stabilized zirconia, *Journal of the American Ceramic Society*, **73**, 893 – 903.

Davidenkov, N. (1947) The influence of size on the brittle strength of steel, *Journal of Applied Mechanics*, **14**, 63 – 7.

Davidge, R.W. (1989) The mechanical properties and fracture of ceramic matrix composites (CMC) reinforced with continuous fibres, in *Application of Fracture Mechanics to Composite Materials* (ed. K. Friedrich), Elsevier Science Publishers, Amsterdam, pp. 547 – 69.

Davidson, D.L. (1992) The experimental mechanics of microcracks, in *Small-Crack Test Methods* (eds J.M. Larsen and J.E. Allison), ASTM STP 1149, American Society for Testing and Materials, Philadelphia, pp. 81 – 91.

Davidson, D.L. (1993) Fatigue crack growth through composites with continuous fiber reinforcement, in *Proceedings 9th International Conference on Composite Materials* (ed. A. Miravete), Woodhead Publishing, pp. 571 – 7.

Davies, P. and Benzeggagh, M.L. (1989) Interlaminar mode I fracture testing, in *Application of Fracture Mechanics to Composite Materials* (ed. K. Friedrich), Elsevier Science Publishers, Amsterdam, pp. 81 – 112.

Devlin, P.W. (1990) Disharmony in the European Community's product liability laws: is the United Kingdom off-key?, *Temple International and Comparative Law Journal*, **4**, 133.

Donaldson, D.R. and Anderson, W.E. (1962) Crack propagation behavior of some aircraft materials, in *Proceedings Crack Propagation Symposium*, Vol. II, Cranfield College of Aeronautics, Cranfield, UK, pp. 375 – 441.

Dowling, N.E. (1977) Crack growth during low cycle fatigue of smooth axial specimens, in *Cyclic Stress, Strain and Plastic Deformation Aspects of Fatigue Crack Growth* (ed. L.F. Impellizzeri), ASTM STP 637, American Society for Testing and Materials, Philadelphia, pp. 97 – 121.

Dvorak, G.J., (ed.) (1990) *Inelastic Deformation of Composite Materials*, Springer-Verlag, Berlin.

Edwards, P.R. and Newman, J.C. Jr. (1990) An AGARD supplemental test programme on the behavior of short cracks under constant amplitude and aircraft spectrum loading, *in Short-Crack Growth in Various Aircraft Materials*, AGARD Report 767.

Elber, W. (1971) The significance of fatigue crack closure, in *Symposium on Damage Tolerance in Aircraft Structures*, ASTM STP 486, American Society for Testing and Materials, Philadelphia, pp. 230 – 242.

Epstein, B. (1948) Statistical aspects of fracture problems, *Journal of Applied Physics*, **19**, 140 – 147.

Erdogan, F. and Sih, G.C. (1963) On the crack extension in plates under plane loading and transverse shear, *Trans. ASME, Journal of Basic Engineering*, **85**, 519 – 27.

Eshelby H.D. (1969) The starting of a crack, in *Physics of Strength and Plasticity* (ed. A.S. Argon), MIT Press, Cambridge, Mass., pp. 263 – 75.

Evans, A.G. (1980) Fatigue in ceramics, *International Journal of Fracture*, **16**, 485 – 98.

Evans, F.G. and Lebow, M. (1957) Strength of human compact bone under repetitive loading, *Journal of Applied Physiology*, **10**, 127 – 30.

Evans, F.G. and Riolo, M.L. (1970) Relations between the fatigue life and histology of adult human cortical bone, *Journal of Bone Joint Surgery*, **53-A**, 1579 – 86.

Everett, R.A., Bartlett, F.O. and Elber, W. (1990) *Probabilistic Fatigue Methodology for Six Nines Probability*, NASA Technical Memorandum 102757, AVSCOM Technical Report 90-8-009, Hampton, Va.

Ewart, K. and Suresh, S. (1987) Crack propagation in ceramics under cyclic loads, *Journal of Materials Science*, **22**, 1173 – 92.

Ewart, L. (1990) *Ambient and Elevated Temperature Fatigue Crack Growth in Alumina*, Ph.D. Thesis, Brown University, Providence, RI.

Fagan, B. (1992) *People of the Earth*, 7th edn, Harper Collins, New York.

Fairburn, W. (1864) Experiments to determine the effect of impact, vibratory action, and long continued changes of load on wrought iron girders, *Philosophical Transactions of the Royal Society*, **154**, 311.

Faupel, J.H. and Fisher, F.E. (1981) *Engineering Design*, Wiley-Interscience, New York.

Fedderson, C. (1967) Discussion, in *Plane Strain Crack Toughness Testing of High Strength Metallic Materials* (eds W. F. Brown and J. E. Srawley), ASTM STP 410, American Society for Testing and Materials, Philadelphia, p. 77.

Fishman, S.G. (1991) Control of interfacial behavior in structural composites, in *Proceedings 8th International Conference on Composite Materials* (eds S.W. Tsai and G.S Springer), Society for the Advancement of Materials and Process Engineering, Covina, Ca, pp. 19A 1 – 12.

Forrest, P.G. (1962) *Fatigue of Metals*, Pergamon Press, London.

Forsyth, P.J.E. (1953) Extrudation of material from slip bands at the surface of fatigued crystals of an aluminum – copper alloy, *Nature*, **171**, 172 – 3.

Forsyth, P.J.E. (1962) A two stage process of fatigue crack growth, in *Proceedings Crack Propagation Symposium*, Vol. 1, Cranfield College of Aeronautics, Cranfield, UK, pp. 76 – 94.

Forsyth, P.J.E. (1983) A unified description of micro and macroscopic fatigue crack behavior, *International Journal of Fatigue*, **5**, 3 –14.

Freitag, T.A. and Cannon, S.L. (1977) Fracture characteristics of acrylic bone cements II, Fatigue, *Journal of Biomedical Materials Research*, **11**, 602 – 24.

Frost, N.E. (1959) Propagation of fatigue cracks in various sheet materials, *Journal of Mechanical Engineering Science*, **1**, 151 – 70.

Frost, N.E. (1960) Notch effects and the critical alternating stress required to propagate a crack in an aluminum alloy subject to fatigue loading, *Journal of Mechanical Engineering Science*, **2**, 109 – 19.

Frost, N.E. and Dugdale, D.S. (1957) Fatigue tests on notched mild steel plates with measurements of fatigue cracks, *Journal of Mechanics and Physics of Solids*, **5**, 182 – 92.

Frost, N.E., Marsh, K.J. and Pook, L.P. (1974) *Metal Fatigue*, Clarendon Press, Oxford.

Fuchs, H.O. and Stephens, R.I. (1980) *Metal Fatigue in Engineering*, John Wiley & Sons, New York.

Fung, Y.C. (1965) *Foundations of Solid Mechanics*, Prentice-Hall, Englewood Cliffs, NJ.

Gangloff, R.P., Stavik, D.C., Piascik, R.S. *et al.* (1992) Direct current electrical potential measurement of the growth of small cracks, in *Small-Crack Test Methods* (eds J.M. Larsen and J.E. Allison), ASTM STP 1149, American Society for Testing and Materials, Philadelphia, pp. 116 – 68.

Gao, Z. (1992) Modeling and simulation of composites under fatigue loading, in *Theoretical Concepts and Numerical Analysis of Fatigue* (eds A.F. Blom and C.J. Beevers), EMAS, Warley, UK, pp. 465 – 78.

Gao, Z., Reifsnider, L.L. and Carman, G.P. (1992) Strength prediction and optimization of composites with statistical fiber flaw distributions, *Journal of Composite Materials*, **26**, 1698 –1705.

Garvie, R.C., Hannick, R.H. and Pescoe, R.T. (1975) Ceramic steel?, *Nature*, **258**, 703 – 4.

Gerber, H. (1874) Bestimmung der zulässigen Spannungen in Eisenkonstructionen, *Zeitschrift des Bayerischen Architeckten and Ingenieur-Vereins*, **6**, 101 – 10.

Gieseke, B. and Saxena, A. (1989) Correlation of creep fatigue crack growth rates using crack tip parameters, in *Advances in Fracture Research* (eds K. Salama, K. Ravi-Chandar, D.M.R. Taplin and P. Rama Rao) Vol. 1, Pergamon Press, Oxford, pp. 189 – 96.

Goodman, J. (1899) *Mechanics Applied to Engineering*, Longmans-Green, London.

Green, A.E. and Zerna, W. (1968) *Theoretical Elasticity*, 2nd edn, Oxford University Press, Oxford, UK.

Greenfield, I.G., Fang, N.J-J. and Orthlieb, F.L. (1991) Reorientation of SiC whiskers in aluminum at different stages of extrusion, composites, in *Proceedings 8th International Conference on Composite Materials*, (eds S.W. Tsai and G.S. Springer) Society for Advancement of Materials and Process Engineering, Covina, Ca, pp. 17E 2 – 9.

Griffith, A.A. (1920) The phenomenon of rupture and flow in solids, *Philosophical Transactions of the Royal Society*, **221A**, 163 – 98.

Gross, B. and Srawley, J.E. (1972) Stress intensity factors for bend and compact specimens, *Engineering Fracture Mechanics*, **4**, 587 – 9.

Gross, B., Srawley, J.E and Brown, W.E. Jr. (1964) *Stress Intensity Factors for a Single Edge Notch Tension Specimen by a Boundary Collocation of a Stress Function*, NASA Technical Note D-2395.

Guiu, F. (1978) Cyclic fatigue of polycrystalline alumina in direct push – pull. *Journal of Materials Science Letters*, **13**, 1357 – 61.

Gupta, V., Argon, A.S., Cornie, J.A. *et al.* (1992) Measurement of interface strength by a laser spallation technique, *Journal of Mechanics and Physics of Solids*, **40**, 141 – 80.

Guy, A.G. (1962) *Physical Metallurgy for Engineers*, Addison-Wesley, Reading, Mass.

Halliday, M. (1994) Unpublished research at the *University of Birmingham Laboratory for Materials for High Performance Applications*, Birmingham, UK.

Halliday, M.D. and Beevers, C.J. (1981) Some aspects of crack closure in two contrasting titanium alloys, *Journal of Testing and Evaluation*, **9**, 195 – 201.

Han, L.X. and Suresh, S. (1989) High temperature crack growth in an alumina – silicon carbide whisker composite: mechanisms of fatigue tip damage, *Journal of American Ceramic Society*, **72**, 1233 – 8.

Hashin, Z. (1986) Analysis of stiffness reduction of cracked cross-ply laminates, *Engineering Fracture Mechanics*, **25**, 771 – 8.

Hashin, Z. and Rotem, A. (1973) A fatigue failure criterion for fiber reinforced composites, *Journal of Composite Materials*, **7**, 448 – 64.

He, M.Y. and Hutchinson, J.W. (1989) Kinking of a crack out of an interface, *Journal of Applied Mechanics*, **56**, 270 – 8.

Heitz, E., Henkhaus, R. and Rahmel, A. (1992) *Corrosion Science – An Experimental Approach*, Ellis Horwood, Chichester, UK.

Hellan, Kåre (1984) *Introduction to Fracture Mechanics*, McGraw-Hill, New York.

Hertzberg, R.W. (1989) *Deformation and Fracture of Engineering Materials*, 3rd edn, Wiley, New York.

Hertzberg. R.W. and Manson, J.A. (1980) *Fatigue of Engineering Plastics*, Academic Press, New York.

Hertzberg, R.W. and Manson, J.A. (1986) Fatigue, in *Encyclopedia of Polymer Science and Engineering* (ed. J.I. Kroschwitz) Vol. 7, Wiley, New York, pp. 378 – 453.

Hertzberg, R.W., Manson, J.A. and Skibo, M.D. (1975) Frequency sensitivity of fatigue processes in polymeric solids, *Polymers in Engineering and Science*, **15**, 252 – 60.

Hertzberg, R.W., Skibo, M.D. and Manson, J.A. (1979) Fatigue fracture mechanisms in engineering plastics, in *Fatigue Mechanisms* (ed. J.T. Fong), ASTM STP 675, American Society for Testing and Materials, Philadelphia, pp. 471 – 500.

Hertzberg, R., Herman, W.A., Clark, T. *et al.* (1992) Simulation of short crack and other low closure loading conditions utilizing constant $k_{max}$ & decreasing fatigue crack growth procedures, *Small-Crack Test Methods* (eds J.M. Larsen and J.E. Allison), ASTM STP 1149, American Society for Testing and Materials, Philadelphia, pp. 197 – 220.

Heuler, P. and Schütz, W. (1985) Fatigue life prediction in the crack initiation and crack propagation stages, in *Durability and Damage Tolerance in Aircraft Design* (eds A. Salvetti and G. Cavallini), EMAS, Warley, UK, pp. 38 – 69.

Hinkley, J.A. and O'Brien, T.K. (1991) Delamination behavior of quasi-isotropic graphite epoxy laminates subjected to tension and torsion loads, in *Proceedings American Helicopter Society – Rotorcraft Structures Specialists Meeting*, American Helicopter Society, Williamsburg, Va.

Hole, F. and Heizer, R.F. (1965) *An Introduction to Prehistoric Archeology*, Holt, Rinehart & Winston, New York.

Holmes, J.W. (1991) Influence of stress ratio on elevated temperature fatigue of a SiC fiber reinforced $Si_3N_4$ composite, *Journal of American Ceramic Society*, **74**, 1639 – 45.

Howard, W.E., Gossard,T. Jr. and Jones, R.M. (1989) Composite laminate free-edge reinforcement with u-shaped caps, part II: theoretical-experimental correlation, *American Institute of Aeronautics and Astronautics Journal*, **27**, pp. 617 – 23.

Howells, G. (1993) *Comparative Product Liability*, Dartmouth Publ., Brookfield, Vermont.

Huang, Y.H. and Wang, S.S. (1988) Compressive fatigue damage and associated property degradation of aluminum matrix composite, in *Proceedings 4th Japan – U.S. Conference on Composite Materials*, American Society for Composites, Technomic, Lancaster, Pa, pp. 606 – 632.

Hull, D. (1981) *An Introduction to Composite Materials*, Cambridge University Press, Cambridge, UK.

Hurd, S. and Zollers, F. (1993) Product liability in the European Community: implications for United States business, *American Business Law Journal*, **31**, 245.

Hutchinson, J.W. (1983) Fundamentals of the phenomenological theory of nonlinear fracture mechanics, *Journal of Applied Mechanics*, **50**, 1042 – 51.

Hutchinson, J.W. (1987) Crack tip shielding by micro-cracking in brittle solids, *Acta Metallurgica*, **35**, 1605 – 19.

Hutchinson, J.W. (1990) Mixed mode fracture mechanics of interfaces, in *Metal – Ceramic Interfaces* (eds M. Ruhle, A.G. Evans, M.Ashby and J. Hirth), Pergamon Press, New York, p. 307.

Hutchinson, J.W. and Suo, Z. (1992) Mixed mode cracking in layered materials, in *Advances in Applied Mechanics* (eds J.W. Hutchinson and T.Y. Wu), Vol. 29, Academic Press, London, UK, pp. 63 – 191.

Hutchinson, J.W., Mear, M.E. and Rice, J.R. (1987) Crack Paralleling an interface between dissimilar materials, *Transactions American Society of Mechanical Engineers*, **54**, 828 – 832.

Inglis, C.E. (1913) Stresses in a plate due to the presence of cracks and sharp corners, *Proceedings Institute of Naval Architects*, **55**, 219.

Irwin, G.R. (1948) Fracture Dynamics, *Fracturing of Metals*, (ed. G. Sachs) ASM, Cleveland pp, 147 – 66.

Irwin, G.R. (1957) Analysis of stresses and strains near the end of a crack traversing a plate, *Journal of Applied Mechanics*, **24**, 361 – 4.

Johnson, A.F. (1979) *Engineering Design Properties of GRP*, British Plastics Federation, London.

Johnson, L.G. (1964) *The Statistical Treatment of Fatigue Experiments*, Elsevier, London.

Johnson, W. and Mellor, P.B. (1973) *Engineering Plasticity*, Van Nostrand Reinhold, London.

Jones, R. and Calliman, R.J. (1979) Finite element analysis of patched cracks, *Journal of Structural Mechanics*, **7**, 107 – 30.

Jones, R.E. (1973) Fatigue crack growth retardation after single cycle peak overload Ti-6Al-4V titanium alloy, *Engineering Fracture Mechanics*, **5**, 585 – 604.

Jones, R.E. (1975) *Mechanics of Composite Materials*, McGraw-Hill, New York.

Jones, R.M. (1991) Delamination-suppression concepts for composite laminate free edges, in *Proceedings 8th International Conference on Composite Materials* (eds S.W. Tsai and G.S. Springer), Society for Advancement Material and Process Engineering, Covina, Ca, pp. 28M 1 – 10.

Juvinall, R.C. (1967) *Engineering Considerations of Stress, Strain and Strength*, McGraw-Hill, New York.

Kachanov, M. (1988) On modeling of a microcracked zone by a weakened elastic material and statistical aspects of crack – microcrack interactions, *International Journal of Fracture*, **37**, R55 – 62.

Kageyama, K., Kikuchi, M. and Yanagusawa, N. (1991) Stabilized end notch flexure test: characterization of mode II interlaminar crack growth, in *Composite Materials – Fatigue and Fracture* (ed. T.K. O'Brien), ASTM STP 1110, American Society for Testing and Materials, Philadelphia, pp. 210 – 25.

Kambour, R.P. (1964) Structure and properties of crazes in polycarbonate and other glassy polymers, *Polymer*, **5**, 143 – 55.

Kardomateas, G.A. (1993) The initial postbuckling and growth behavior of internal delaminations in composite plates, *Journal of Applied Mechanics*, **60**, 903 – 10.

Kardomateas, G.A. and Carlson, R.L. (1995) An analysis of compressive load excursions on fatigue crack growth in metallic materials, *Journal Applied Mechanics*, **62**, 240 – 3.

Kardomateas, G.A., Carlson, R.L., Soedino, A.H. *et al.* (1993) Near tip stress and strain fields for short elastic cracks, *International Journal of Fracture*, **62**, 219 – 32.

Kardomateas, G.A., Pelegri, A.A. and Malik, B. (1995) Growth of internal delaminations under cyclic compression in composite plates, *Journal of the Mechanics and Physics of Solids*, **43**, 847 – 68.

Kausch, H.H. (1987) *Polymer Fracture*, 2nd edn, Springer-Verlag, Berlin.

Keer, L.M. and Freedman, J.M. (1973) Tensile strip with edge cracks, *International Journal of Engineering Science*, **11**, 1265 – 75.

Keeton, W.P. (ed.) (1984) *Prosser and Keeton on the Law of Torts*, 5th edn,West Publ., St. Paul, Minnesota.

Kelly, A. and Bomford, M.J. (1969) Fatigue of the matrix in a fibre-reinforced composite, in *Physics of Strength and Plasticity* (ed. A.S. Argon), MIT Press, Cambridge, Mass, pp. 339 – 50.

Kemper, H., Weiss, B. and Stickler, R. (1989) An alternative presentation of the effects of the stress-ratio on fatigue threshold, *Engineering Fracture Mechanics*, **4**, 591 – 600.

Kendall, J.M. and Knott, J.F. (1984) The influence of microstructure and temperature on near threshold fatigue crack growth in air and vacuum, in *Proceedings 2nd International Conference on Fatigue and Fatigue Thresholds* (ed. C.J. Beevers), EMAS, Warley, UK, pp. 307 – 17.

Kingery, W.D. (1960) *Introduction to Ceramics*, John Wiley, New York.

Kinloch, A.J. and Young, R.J. (1983) *Fracture Behavior of Polymers*, Applied Science, London.

Kirchner, H.P., Gruver, R.M. and Walker, R.E. (1968) Strengthening of alumina by glazing and quenching, *American Ceramic Society Bulletin*, **47**, 798 – 802.

Kishimoto, H., Veno, A., Kawamoto, H., *et al.* (1992) Fatigue crack propagation behavior and its mechanism in alumina, in*Theoretical Concepts and Numerical Analysis of Fatigue* (eds A.F. Blom and C.J. Beevers), EMAS, Warley, UK, pp. 559 – 570.

Kitagawa, H. and Takahashi, S. (1976) Applicability of fracture mechanics to very small cracks or cracks in the early stage, in *Proceedings 2nd International Conference on Mechanical Behavior of Materials*, American Society for Metals, Metals Park, Oh, pp. 627 – 31.

Knott, J.F. (1973) *Fundamentals of Fracture Mechanics*, Butterworths, London.

Knott, J.F. (1986) Models of fatigue crack growth, in *Fatigue Crack Growth - 30 Years of Progress* (ed. R.A.Smith), Pergamon Press, Oxford, pp. 31 – 52.

Kochendörfer, R. (1993) Liquid silicon infiltration – a fast and low cost cmc – manufacturing process, in *Proceedings 9th International Conference on Composite Materials* (ed. A. Miravete), Woodhead Publ., Cambridge, UK, pp. 23F 2 –9.

Kreider, K.G. (1974) Introduction to metal-matrix composites, in *Composite Materials*, Vol. 4 (ed. K.G. Kreider), Academic Press, London, pp. 1 – 35.

Kumai, S., King, J.E., and Knott, J.F. (1990) Fatigue crack growth in sic particulate reinforced aluminum alloy, in *Proceedings 4th International Conference on Fatigue and Fatigue Thresholds* (eds H. Kitagawa and T. Tanaka), MCEP, Birmingham, UK, pp. 869 – 74.

Laird, C. and Smith, G.C. (1962) Crack propagation in high stress fatigue, *Philosophical Magazine*, **7**, 847 – 57.

Lamon, J. (1992) Probabilistic modeling of damage in brittle materials and ceramic matrix composites, in *Theoretical Concepts and Numerical Analysis of Fatigue* (eds. A.F. Blom and C.J. Beevers), EMAS, Warley, UK, pp. 479 – 504.

Landes, J.D. and Begley, J.A. (1976) A fracture mechanics approach to creep crack growth, in *Mechanics of Crack Growth*, ASTM STP 590, American Society Testing and Materials, Philadelphia, pp. 128 – 48.

Lankford, J. (1982) The growth of small fatigue cracks in 7075-T6 aluminum, *Fatigue of Engineering Materials and Structures*, **5**, 233 – 48.

Lankford, J. and Davidson, D.L. (1982) The effect of overloads upon fatigue crack tip opening displacement & crack tip opening/closing loads in aluminum alloys, in *Advances in Fracture Research* , Vol. 2 (ed. D. Francois), Pergamon Press, Oxford, pp. 899 – 906.

Larsen, J. and Allison, J.E. (eds.) (1992) *Small-Crack Test Methods*, ASTM STP 1149, American Society for Testing and Materials, Philadelphia.

Larsen, J.M., Jira, J.R. and Ravichandran, K.S. (1992) Measurement of small cracks by photomicroscopy: experiments and analysis, in *Small-Crack Test Methods* (eds J.M. Larsen and J.E. Allison), ASTM STP 1149, American Society for Testing and Materials, Philadelphia, pp. 57 – 80.

Larsson, S.G. and Carlsson, A.J. (1973) Influence of nonsingular stress terms and specimen geometry on small scale yielding at crack tip in elastic – plastic solids, *Journal of Mechancis of Physics and Solids*, **21**, 263 – 77.

Leevers, P.S. and Radon, J.C. (1982) Inherent stress biaxiality in various fracture specimen geometries, *International Journal of Fracture*, **9**, 311 – 24.

Leong, K.H. and King, J.E. (1991) Damage accumulation in a $[0,90]_{25}$ GFRP under static and fatigue loading, in *Proceedings 8th International Conference on Composite Materials* (eds E.W. Tsai and G.S. Springer), Society for Advancement of Materials and Process Engineering, Covina, Ca, pp. 38D 1 – 10.

Liaw, P.K., Leax, T.R., Williams, R.S. *et al.* (1982) Near threshold fatigue crack growth behavior in copper, *Metallurgical Transactions*, **13A**, 1607 – 18.

Liechti, K.M. and Chai, Y.S. (1992) Asymmetric shielding in interfacial fracture under in-plane shear, *Journal of Applied Mechanics,*, **59**, 295 – 304.

Lincoln, J.W. (1985) Risk assessment of an aging military aircraft, *Journal of Aircraft*, **22**, 687 – 91.

Lindley, T.C. (1992) Fretting fatigue characteristics and integrity assessment, in *Theoretical Concepts and Numerical Analysis of Fatigue* (eds A.F. Blom and J.C. Beevers), EMAS, Warley, UK, pp. 73 – 84.

Linnig, W. (1993) Some aspects of the prediction of fatigue crack paths, in *Mixed-Mode Fatigue and Fracture* (eds H.P. Rossmanith and K.J. Miller), Mechical Engineering Publ, London, pp. 201 – 15.

Lipson, C. and Juvinall, R. (1963) *Handbook of Stress and Strength*, Macmillan, New York.

Liu, H.W. (1963) Fatigue crack propagation and applied stress range – an energy approach, *Journal of Basic Engineering*, **85**, 116 – 22.

Lukás, P. and Klesnil, M. (1978) Fatigue limit of notched bodies, *Materials, Science and Engineering*, **34**, 61 – 6.

Mandell, J.F., McGarry, F.J., Huang, D.D. *et al.* (1983) Some effects of matrix and interface properties on the fatigue of short fiber-reinforced thermoplastics, *Polymer Composites*, **4**, 32 – 9.

Manning, S.D. and Yang, J.N. (1987) *Advanced Durability Analysis, Vol.I – Analytical Methods*, AFWAL-TR-86-3017, Flight Dynamics Lab., Wright-Patterson Air Force Base, Oh.

Manning, S.D., Yang, J.N. and Welch, K.M. (1992) Aircraft structural maintenance scheduling based on risk and individual aircraft tracking, in *Theoretical Concepts and Numerical Analysis of Fatigue* (eds A.F. Blom and C.J. Beevers), EMAS, Warley, UK, pp. 401 – 20.

Manson, S.S. (1954) *Behavior of Materials Under Conditions of Thermal Stress*, NACA: Report 1170, Lewis Flight Propulsion Laboratory, Cleveland, Oh.

Manson, S.S. and Hirschberg, M.H. (1964) *Fatigue: An Interdisciplinary Approach*, Syracuse University Press, Syracuse, NY, p. 133.

Marom, G. (1989) Environmental effects on fracture mechanical properties of polymer composites, in *Application of Fracture Mechanics to Composite Materials* (ed. K. Friedrich), Elsevier Science Publishers, Amsterdam, pp. 39 – 42.

Marsh, K.J. and Smith. R.A. (1986) Thirty years of fatigue testing equipment, in *Fatigue Crack Growth – 30 Years of Progress* (ed. R.A. Smith), Pergamon Press, Oxford, pp. 17 – 30.

Marshall, D.B. (1986) Strength characteristics of transformation toughened zirconia, *Journal of American Ceramic Society*, **69**, 173 – 80.

Marshall, D.B. and Cox, B.N. (1987) Tensile fracture in brittle matrix composites: influence of fiber strength, *Acta Metallurgica*, **35**, 2607 – 19.

Martin, R.H. and Murri, G.B. (1990) Characterization of mode I and mode II delamination growth and thresholds in AS4/PEEK composites, *in Composite Materials: Testing and Design* (ed. S.P. Garbo), ASTM STP 1059, American Society for Testing and Materials, Philadelphia, pp. 251 – 270.

McCartney, L.M. (1987) Mechanics of matrix cracking in brittle-matrix fibre-reinforced composites, *in Proceedings of the Royal Society*, London, **A409**, 329 – 50.

McClintock, F.A. (1955) A criterion for minimum scatter in fatigue testing, *Journal of Applied Mechanics*, **22**, 427 – 31.

McClintock, F.A. (1963) On the plasticity of the growth of fatigue cracks, in *Fracture of Solids,* Vol. 20 (eds D.C. Drucker and J.J. Gilman), Wiley, New York, pp. 65 – 102

McClintock, F.A. (1972) Fracture mechanics for corrosion fatigue, in *Proceedings of the Conference on Corrosion Fatigue: Chemistry, Mechanics and Microstructure* (eds O. Devereux, A.J. McEvily and R.W. Staehle), National Association of Corrosion Engineers, Houston, pp. 289 – 302.

McClintock, F.A. and Argon, A.S. (eds) (1966) *Mechanical Behavior of Materials*, Addison-Wesley, Reading, Mass.

McClung, R.C. and Sehitoglu, H. (1988) Closure behavior of small cracks under high strain fatigue histories, in *Mechanics of Fatigue Crack Closure* (eds J.C. Newman, Jr. and W. Elber), ASTM STP 982, American Society for Testing and Materials, Philadelphia, pp. 279 – 99.

McClung, R.C. (1992) Finite element modeling of fatigue crack growth, in *Theoretical Concepts and Numerical Analysis of Fatigue* (eds A.F. Bloom and C.J. Beevers), EMAS, Warley, UK, pp. 153 – 72.

McEvily, A.J. and Wei, R.P. (1972) Fracture mechanics and corrosion fatigue, in *Proccedings of the Conference on Corrosion-Fatigue: Chemistry, Mechanics and Microstructure* (eds O.F. Devereux, A.J. McEvily and R.W. Staehle), National Association of Corrosion Engineers, Houston, pp. 381 – 95.

McEvily, A.J. and Yang, Z. (1990) Fatigue crack growth retardation mechanisms of single and multiple overloads, in *Proceedings 4th International Conference on Fatigue and Fatigue Thresholds* (eds H. Kitagawa and T. Tanaka), MCEP, Birmingham, UK, pp. 23 – 36.

McGonnagle, W.J. (1961) *Nondestructive Testing*, McGraw-Hill, New York.

McMaster, R.C. (1959) *Non-destructive Testing Handbook*, Vols I and II, edited for American Society for Non-destructive Testing, Ronald Press, New York.

McMeeking, R.M. and Evans, A.C. (1982) Mechanics of transformation toughening in brittle materials, *Journal of American Ceramic Society*, **65**, 242 – 5.

Merkle, J.G. and Corten, H.T. (1974) A *J* integral analysis of the compact specimen considering axial force as well as bending effects, *Journal of Pressure Vessel Technology*, **98**, 28, 92.

Metcalf, A.G. (1974) Fiber-reinforced titanium alloys, in *Composite Materials*, Vol. 4 (ed. K.G. Kreider), Academic Press, London, pp. 269– 318.

Mignery, L.A., Tan, T.M. and, Sun C.T. (1985) The use of stitching to suppress delamination in laminated composites, in *Delamination and Debonding* (ed. W.S.Johnson), ASTM STP 876, American Society for Testing and Materials, Philadelphia, pp. 371 – 85.

*MIL Handbook 5*, Department of Defense, Washington, DC.

Miller, K.J. and de los Rios, E.R.(eds) (1986) *The Behavior of Short Fatigue Cracks*, Mechanical Engineering Publ., London.

Miner, M.A. (1945) Cumulative damage in fatigue, *Journal of Applied Mechanics*, **67**, A 159 – 64.

Moftakhar, A. and Glinka, G. (1992) Elastic – plastic stress – strain analysis methods for notched bodies, in *Theoretical Concepts and Numerical Analysis of Fatigue* (eds A.F. Blom and C.J. Beevers), EMAS, Warley, UK, pp. 327 – 41.

Moore, H.F. and Kommers, J.B. (1927) *The Fatigue of Metals*, McGraw-Hill, New York.

Morris, J.M. and Blickenstaff, L.D. (1967) *Fatigue Fractures*, Thomas, Springfield, Ill.

Morris, W.L. (1979) Microcrack closure phenomena for Al 2219-T851, *Metallurgical Transactions*, **10A**, 5 –11.

Morrison, T.W. (1955) Some unusual conditions encountered in lubrication of rolling contact bearings, *Lubrication Engineering*, **11**, 405 – 11.

Murakami, Y. and Endo, M. (1992) The area parameter model for small defects and nonmetallic inclusions in fatigue strength: experimental evidence and applications, in *Theoretical Concepts and Numerical Analysis of Fatigue* (eds A.F. Blom and C.J. Beevers), EMAS, Warley, UK, pp. 51 – 71.

Muskhelishvili, N.I. (1963) *Mathematical Theory of Elasticity*, 3rd edn, Noordhoff, Groningen.

Nadai, A. (1931) *Plasticity*, McGraw-Hill, New York.

NASA/FLAGRO (1989) *Fatigue Crack Growth Computer Program*, NASA, Lyndon B. Johnson Space Center, Houston, Texas, p. 5.

Neuber, H. (1946) *Theory of Notch Stresses: Principle for Exact Stress Calculations*, Edwards, Ann Arbor, Mich.

Neuber, H. (1961) Theory of stress concentrations for shear – strained prismatical bodies with arbitrary nonlinear stress – strain laws, *Journal of Applied Mechanics*, **E28**, 544 – 51.

Newman, J.C. Jr. (1971) *An Improved Method of Collocation for the Stress Analysis of Cracked Plates with Various Shaped Boundaries*, NASA Technical Note D-6376, Langley, Va.

Newman, J.C. Jr. (1981) A crack closure model for predicting fatigue crack growth under aircraft spectrum loading, in *Methods and Models for Predicting Fatigue Crack Growth under Random Loading* (eds J.B. Chang and C.M. Hudson), ASTM STP 748, American Society for Testing and Materials, Philadelphia, pp. 53 – 84.

Newman, J.C. Jr. (1984) A crack opening stress equation for fatigue crack growth, *International Journal of Fracture*, **24**, R131 – 5.

Newman, J.C. Jr. (1992a) FASTRAN II *A Fatigue Crack Growth Structural Analysis Program*, NASA TM 104159.

Newman, J.C. Jr. (1992b) Fracture mechanics parameters for small fatigue cracks, in *Small-Crack Test Methods* (eds. J.E. Larsen and J.E. Allison), ASTM STP 1149, American Society for Testing and Materials, Philadelphia, pp. 6 – 33.

Nisitani, H., Goto, M. and Kawagoishi, N. (1990) Significance of small-crack growth law in fatigue behavior of materials, in *Proceedings 4th International Conference on Fatigue and Fatigue Thresholds* (eds H. Kitagawa and T. Tahaka), MCEP, Birmingham, UK, pp. 1067 – 72.

Norton, F.N. (1974) *Elements of Ceramics*, 2nd edn, Addison-Wesley, Reading, Mass.

Nowack, H., Trautmann, K.H., Schulte, K. *et al.* (1979) Sequence effects on fatigue crack propagation: mechanical and microstructural contributions, in *Fracture Mechanics* (ed. C.W. Smith) ASTM STP 677, American Society for Testing and Materials, Philadelphia, pp. 36 – 53.

O'Brien, T.K. (1982) Characterization of delamination onset and growth in a composite laminate, in *Damage in Composite Materials* (eds K.L. Reifsnider and W.S. Johnson), ASTM STP 775, American Society for Testing and Materials, Philadelphia, pp. 140 – 67.

O'Brien, T.K. (1985) Analysis of local delaminations and their influence on composite laminate behavior, in *Delamination and Debonding of Materials* (ed. W.S. Johnson), ASTM STP 876, American Society for Testing and Materials, Philadelphia, pp. 282 – 97.

O'Brien, T.K. (1990) Towards a damage tolerance philosophy for composite materials and structures, in *Composite Materials: Testing and Design* Vol. 9 (ed. S.P.Garbo), ASTM STP 1059, American Society for Testing and Materials, Philadelphia, pp. 7 – 33.

O'Brien, T.K. (1991) Delamination, durability and damage tolerance of laminated composite materials, in *Proceedings 8th International Conference on Composite Materials* (eds. S.W. Tsai and G.S. Springer), Society for the Advancement of Materials and Process Engineering, Covina, Ca, pp. 28 A 2 – 13.

Ochiai, S. (1989) Fracture mechanical approach to metal matrix composites, in *Application of Fracture Mechanics to Composite Materials* (ed. K. Friedrich), Elsevier Science Publishers, Amsterdam, pp. 491 – 545.

O'Conner, P.D.T. (1988) *Reliability Engineering*, Hemisphere Publ., New York.

Ogawa, T., Tokunori, O. and Keiro, T. (1993) On the relationship between fatigue crack growth and fracture resistance curve in ceramics, in *Proceedings 9th International Conference on Composite Materials* (ed. A. Miravete), Woodhead Publ., Cambridge, UK, pp. 129 – 136.

Orowan, E. (1952) Fundamentals of brittle behavior in metals, in *Fatigue and Fracture of Metals* (ed. W.M. Murray), Wiley, New York, pp. 139 – 67.

Ortiz, M. (1987) A continuum theory of crack shielding in ceramics, *Journal of Applied Mechanics*, **54**, 54 – 8.

Ortiz, M. (1988) Microcrack coalescence and macroscopic crack growth initiation in brittle solids, *International Journal of Solids and Structures*, **24**, 231 – 50.

Osgood, C.C. (1982) *Fatigue Design*, 2nd edn, Pergamon Press, Oxford.

Owen, C.R., Bucci, R.J. and Kegarise, R.J. (1989) An aluminum quality breakthrough for aircraft structural reliability, *Journal of Aircraft*, **26**, 178 – 84.

Owen, D.R.J. and Fawkes, A.J. (1983) *Engineering Fracture Mechanics: Numerical Methods and Applications*, Pineridge Press, Swansea, UK.

Pagano, N.J. and Whitney, J.M. (1970) Geometric design of composite cylindrical characterization specimens, *Journal of Composite Materials*, **4**, 360 – 78.

Pagano, N.J. and Pipes, R.B. (1971) The influence of stacking sequence on laminate strength, *Journal of Composite Materials*, **5**, 50 – 7.

Palaniswany, K. and Knauss, W.G. (1978) On the problem of crack extension in brittle solids under general loading, in *Mechanics Today* (ed. S. Nemat-Nasser), Pergamon Press, New York, pp. 87 – 148.

Palmgren, A. (1924) Die Lebensdauer von Kugellagern, 2, *Verein Deutscher Ingenieure*, **68**, 339 – 47.

Paris, P.C. and Erdogan, F. (1963) Critical analysis of crack propagation laws, *Trans. ASME, Journal of Basic Engineering*, **85**, 528 – 34.

Paris, P.C. and Hermann, L. (1982) Twenty years of reflection on questions involving fatigue crack growth, part II: some observations on crack closure, in *Proceedings 1st International Conference on Fatigue Thresholds* (eds J. Backlund, A.F. Blom and C.J. Beevers) EMAS, Warley, UK pp. 11 – 33.

Paris, P.C., Gomez, M.P. and Anderson, W.E. (1961) A rational analytic theory of fatigue, in *The Trend in Engineering*, **13**, 9 – 14.

Paris, P.C., Bucci, R.J., Wessel, E.T. *et al.* (1972) Extensive study of low fatigue crack growth rates in A533 and A508 steels, in *Stress Analysis and Growth of Cracks*, ASTM STP 513, American Society for Testing and Materials, Philadelphia, pp. 141 – 76.

Park, J.H. and Atluri, S.N. (1993) Fatigue growth of multiple cracks near a row of fastener holed in a fuselage lap joint, *Computational Mechanics*, **13**, 189 – 203.

Park, J.H., Ogiso, T. and Atluri, S.N. (1992) Analysis of cracks in aging aircraft structures with and without composite patch repairs, *Computational Mechanics*, **10**, 169 -202.

Park, J.H., Singh, R., Pyo, C.R *et al.* (1993) Integrity of aircraft structural elements with multi-site fatigue damage, in *Durability and Structural Integrity of Airframes* (ed. A.F. Blom) EMAS, Warley, UK, pp. 1 – 42.

Parker, E.R. (1957) *Brittle Behavior of Engineering Structures*, John Wiley, New York.

Partl, O. and Schijve, J. (1990) Reconstruction of crack growth from fractographic observations after flight simulation loading, *International Journal of Fatigue*, **12**, 175 – 83.

Pearson, S. (1975) Initiation of fatigue cracks in commercial aluminum alloys and the subsequent propagation of very short cracks, *Engineering Fracture Mechanics*, **7**, 235 – 47.

Peterson, R.E. (1950) Interpretation of service fractures, in *Handbook of Experimental Stress Analysis* (ed. M. Hetényi), Wiley, New York, pp. 593 – 635.

Peterson, R.E. (1953) *Stress Concentration Design Factors*, John Wiley, New York.

Peterson, R.E. (1959) *Analytical Approach to Stress Concentration Effects in Aircraft Materials*, Technical Report 59-507, US Air Force – WADC Symposium on Fatigue of Metals, Dayton, Oh.

Pickard, A.C. (1985) *The Application of 3-Dimensional Finite Element to Fracture Mechanics and Fatigue Life Prediction*, EMAS, Warley, UK.

Pickard, A.C., Brown, C.W. and Hicks, M.A. (1983) The development of advanced specimen testing and analysis techniques applied to fracture mechanics lifing of gas turbine components, in *Advances in Life Prediction Methods* (eds D.A. Woodward and J.R. Whitehead), ASME, New York, pp. 173-8.

Pindera, M.J., Choksi, G., Hidde, J.S. *et al.* (1987) A methodology for accurate shear characterization of unidirectional composites, *Journal of Composite Materials*, **21**, 1164 – 84.

Pipes, R.B. and Pagano, N.J. (1970) Interlaminar stresses in composite laminates under axial extension, *Journal of Composite Materials*, **4**, 538 – 48.

Pompetzki, M., Topper, T.H. and DuQuesnay, D.L. (1990) The effect of compressive underloads and tensile overloads on fatigue damage accumulation in SAE 1045 steel, *International Journal of Fatigue*, **12**, 207 – 13.

Pook, L.P. (1983) *The Role of Crack Growth in Metal Fatigue*, The Metals Society, London.

Pook, L.P. (1989) The significance of mode I branch cracks for a mixed mode fatigue crack growth threshold behavior, in *Biaxial and Multiaxial Fatigue* (eds M.W. Brown and K.J. Miller), Mechanical Engineering Publ., London pp. 247 – 63.

Poole, P. and Young, A. (1992) Prediction of performance of bonded patch repairs to cracked structures, in *Theoretical Concepts and Numerical Analysis of Fatigue* (eds A.F. Blom and C.J. Beevers), EMAS, Warley, UK, pp. 421 – 38.

Purushothaman, S. and Tien, J.K. (1975) A fatigue crack growth mechanism for ductile materials, *Scripta Metallurgica*, **9**, 923 – 6.

Raj, R. and Baik, S. (1980) Creep crack propagation by cavitation near crack tips, *Metal Science*, **14**, 385 – 93.

Ravichandran, K.S. and Larsen, J.M. (1992) Growth behavior of small and large fatigue cracks in Ti-24 Al-11Nb: effects of crack shape, microstructure and closure, in *Fracture Mechanics* (eds H.A. Ernst, A. Saxena, and D.L. McDowell), ASTM STP 1131, Vol. 1, American Society for Testing and Materials, Philadelphia, pp. 727 – 48.

Reece, M.J., Guiu, F. and Sammur, M.F.R. (1989) Cyclic fatigue propagation in alumina under direct tension-compressive loading, *Journal of the American Ceramic Society*, **72**, 348 – 52.

Reeder, J.R. and Crews, J.H. Jr. (1991) Redesign of the mixed-mode bending test for delamination toughness, in *Proceedings 8th International Conference on Composite Materials*, (eds S.W. Tsai and G.S. Springer), Society for the Advancement of Materials and Process Engineering, Covina, Ca, pp. 36B 1 – 10.

Reifsnider, K.L., Schulte, K. and Duke, J.C. (1983) Long term fatigue behavior of composites, in *Long-term Behavior of Composites,* (ed. T.K. O'Brien), ASTM STP 813, American Society for Testing and Materials, Philadelphia, pp. 136 – 59.

Resch, M.T. and Nelson, D.V. (1992) An ultrasonic method for measurement of size and opening behavior of small fatigue cracks, in *Small-Crack Test Methods* (eds J.M. Larsen and J.E. Allison) ASTM STP 1149, American Society for Testing and Materials, Philadelphia, pp. 169-96.

Restatement of Agency (Second) (1957) sections 343, 344, 350 and 439, American Law Institute.

Rice, J.R. (1967) Mechanics of crack tip deformation and extension by fatigue, in *Fatigue Crack Propagation*, ASTM STP 415, American Society for Testing and Materials, Philadelphia, pp. 247 – 309.

Rice, J.R. (1968) A path independent integral and the approximate analysis of strain concentrations by notches and cracks, *Journal of Applied Mechanics*, **35**, 379 – 86.

Rice, J.R. (1974) Limitations to the small scale yielding approximation for crack tip plasticity, *Journal of Mechanics and Physics of Solids*, **22**, 17 – 26.

Rice, J.R. (1985) First order variations in elastic fields due to variation in location of a planar crack front, *Journal of Applied Mechanics*, **53**, 571 – 9.

Rice, J.R., Paris, P.C. and Merkle, J.G. (1973) Some further results on *J*-integral analysis and estimates, in *Progress in Flaw Growth and Fracture Toughness Testing* (ed. J.G. Kaufman), ASTM STP 536, American Society for Testing and Materials, Philadelphia, pp. 231 – 245.

Rimnac, C.M., Wright, T.M., Bartel, D.L. *et al.* (1986) Failure analysis of a total hip femoral component: a fracture mechanics approach, in *Case Histories Involving Fatigue and Fracture Mechanics* (eds C.M. Hudson and T.P. Rich), ASTM STP 918, American Society for Testing and Materials, Philadelphia, pp. 377 – 88.

Ritchie, R.O. and Knott, J.F. (1973) Mechanisms of fatigue crack growth in low alloy steel, *Acta Metallurgica*, **21**, 639 – 48.

Ritchie, R.O., Suresh, S. and Moss, C.M. (1980) Near threshold fatigue crack growth in 2 1/4 Cr-1Mo pressure vessel steel in air and hydrogen, *Journal of Engineering Materials and Technology*, **102**, 293 – 9.

Ritchie, R.O. and Suresh, S. (1981) Some considerations on fatigue crack closure at near threshold stress intensities due to fracture surface morphology, *Metallurgical Transactions*, **13A**, 937 – 40.

Ritchie, R.O. and Lankford, J. (eds) (1986) *Small Fatigue Cracks*, The Metallurgical Society, Warrendale, Pa.

Ritchie, R.O. and Dauskardt, R.H. (1990) Cyclic fatigue crack propagation behavior in pyrolytic carbon-coated graphites for prosthetic heart valve applications, in *Proceedings 4th International Conference on Fatigue and Fatigue Thresholds* (eds H. Kitagawa and T. Tanaka), MCEP Ltd, Birmingham, UK, pp. 819 – 26.

Rodel, J., Fuller, E.R. and Lawn, B.R. (1991) In situ observations of toughening processes in alumina reinforced with silicon carbide whiskers, *Journal of the American Ceramic Society*, **74**, 3154 – 57.

Rooke, D.R. and Cartwright, D.J. (1976) *Compendum of Stress Intensity Factors*, Hillingdon Press, Oxbridge, UK.

Rose, Eric (1989), International contractual developments in product liability, *Practising Law Institute, Litigation and Administrative Practice Course Handbook Series*, PLI Order No. H-4-5064.

Rose, L.R.F. (1982) A cracked plate repaired by bonded reinforcements, *International Journal of Fracture*, **18**, 135 – 44.

Rosen, B.W. (1972) A simple procedure for experimental determination of the longitudinal shear modulus of unidirectional composites, *Journal of Composite Materials*, **6**, 552 – 4.

Rossmanith, H.P. and Miller, K.J. (eds) (1993) *Proceedings of International Conference on Mixed Mode Fatigue and Fracture*, Mechanical Engineering Publ., London.

Rotem, A. and Hashin, S. (1975) Failure modes of angle-ply laminates, *Journal of Composite Materials*, **9**, 191 – 206.

Russell, A.J. and Street, K.N. (1987) The effect of matrix toughness on delamination static and fatigue fracture under mode II shear loading of graphite fiber composites, in *Toughened Composites* (ed. N.J. Johnston), ASTM STP 937, American Sociey for Testing and Materials, Philadelphia, pp. 275 – 94.

Russell, A.J. and Street, K.N. (1988) A constant $\Delta G$ test for measuring mode I interlaminar fatigue crack growth rates, in *Composite Materials: Testing and Design (8th Conference)* (ed. J.D. Whitcomb), ASTM STP 972, American Society for Testing and Materials, Phildelphia, pp. 259 – 77.

Ryder, J.T. and Crossman, F.W. (1983) *A Study of Stiffness, Residual Strength and Fatigue Life Relationships for Composite Laminates*, NASA Report No. CR-172211, US Department of Commerce National Technical Information Service N84-29978.

Salmassy, O.K., Duckworth, W.H. and Schwope, A.D. (1955a) *Behavior of Brittle State Materials*, Technical Report 53-50, Vol. 1, Wright Air Development Center, Wright-Patterson Air Force Base, Dayton, Oh.

Salmassy, O.K., Bodine, E.G. and Manning, G.K. (1955 b) *Behavior of Brittle State Materials*, Technical Report 53- 50, Vol. 2, Wright Air Development Center, Wright-Patterson Air Force Base, Dayton, Oh.

Sauer, J.A. and Richardson, G.C. (1980) Fatigue of polymers, *International Journal of Fracture*, **6**, 499 – 32.

Savin, G.N. (1961) *Stress Concentration Around Holes*, Pergamon Press, New York.

Saxena, A. (1988) A model for predicting the effect of frequency on fatigue crack growth at elevated temperature, *Fatigue of Engineering Materials and Structures*, **3**, 247 – 55.

Sbaizero, O., Charalambides, P.G., Parros, G., *et al.* (1990) Delamination cracking in a laminated ceramic-matrix composite, *Journal of the American Ceramic Society*, **73**, 1936 – 40.

Schijve, J. (1960) *Fatigue Crack Propagation in Light Alloy Sheet Materials and Structures*, NRL Report MP 195, National Aeronautic and Astronautic Research Institute, Amsterdam.

Schijve, J. (1992) *Multiple-Site-Damage Fatigue of Riveted Joints*, Delft University of Technology, Faculty of Aerospace Engineering Report LR-679, Delft, The Netherlands.

Schijve, J. (1993) Development of fiber-metal laminates (ARALL & GLARE), new fatigue resistant materials, in *Proceedings 5th International Conference on Fatigue and Fatigue Thresholds*, Vol. 1 (eds J.P. Bailon and J.I. Dickson), EMAS, Warley, UK, pp. 3 – 20.

Schijve, J., Van Lipzig, H., Van Gestel, G. *et al.* (1979) Fatigue properties of adhesive-bonded laminated sheet material of aluminum alloys, *Engineering Fracture Mechanics*, **12**, 561 – 79.

Schultz, R.W. and Reifsnider, K.L. (1984) Fatigue of composite laminates under mixed tension-compression loading, in *Proceedings of 2nd International Conference on Fatigue and Fatigue Thresholds* (ed. C.J. Beevers) EMAS, Warley, UK, pp. 1102 – 12.

Scott, P. and Cottis, R.A. (eds) (1990) *Environment Assisted Fatigue*, Mechanical Engineering Publ., London.

Seaman, L.C.B. (1982) *A New History of England 410 – 975*, Harvester Press, Brighton.

Seeger, T. and Heuler, P. (1980) Generalized application of Neuber's rule, *Journal of Testing and Evaluation*, **8**, 199 – 204.

Sela, N. and Ishai, O. (1989) Interlaminar fracture toughness and toughening of laminated composite materials: a review, *Composites*, **20**, 423 – 435.

Sensmeier, M.D. and Wright, P.K. (1990) The effect of fiber bridging on fatigue crack growth in titanium matrix composites, in *Fundamental Relationships Between Microstructure and Mechanical Properties of Metal Matrix Composites* (eds P.K. Liaw and M.N. Gungor), The Minerals, Metals and Materials Society, Warrendale, Pa, pp. 441 – 57.

Shang, J.K., Yu, W. and Ritchie, R.O. (1988) Role of silicon carbide particles in fatigue crack growth in Si C particulate-reinforced aluminum alloy composites, *Materials Science and Engineering*, **102**, 181 – 92.

Sharpe, W.N. Jr., Jira, J.R. and Larsen, J.M. (1992) Real-time measurement of small-crack opening behavior using an interferometric strain/displacement gage, in *Small-Crack Test Methods*, (eds J.M. Larsen and J.E. Allison) ASTM STP 1149, American Society for Testing and Materials, Philadelphia, pp. 92 – 115.

Sheets, P.D. (1993) Dawn of a New Stone Age in eye surgery, in *Archeology: Discovering our Past*, 2nd edn, (eds R.J. Sharer and W. Ashmore), Mayfield, Mountain View, Ca, pp. 470 – 2.

Sheldon, G.P., Cook, T.S., Jones, T.W. *et al.* (1981) Some observations on small fatigue cracks in a superalloy, *Fatigue of Engineering Materials and Structures*, **3**, 319 – 28.

Sheldon, R.P. (1982) *Composite Polymeric Materials*, Applied Science Publ., London.

Shigley, J.E. and Mitchell, L.O. (1983) *Mechanical Engineering Design*, 4th edn, McGraw-Hill, New York.

Shih, C.F. and Hutchinson, J.W. (1976) Fully plastic solutions and large scale yielding estimates for plane stress problems, *Journal of Engineering Materials Technology, Trans. ASME*, **98**, 289 – 95.

Sih, G.C. (1966) On the westergaard method of crack analysis, *Journal of Fracture Mechanics*, **2**, 628 – 31.

Sih, G.C. (1973a) *Handbook of Stress Intensity Factors*, Lehigh University, Bethlehem, Pa.

Sih G.C. (ed.) (1973b) *Methods of Analysis and Solutions to Crack Problems*, Noordhoff, Leyden.

Sih, G.C. (1974) Strain energy density factor applied to mixed mode crack problems, *International Journal of Fracture*, **10**, 305 – 21.

Sih, G.C. and Theocaris, P.S., (eds) (1981) *Mixed Mode Crack Propagation*, Sijthoff & Noordhoff, Rockville, Maryland.

Sih, G.C. (1991) *Mechanics of Fracture Initiation and Propagation – Surface and Volume Energy Density Applied as Failure Criterion*, Kluwer Academic Publ., Dordrecht, the Netherlands.

Sines, G. (1955) *Failure of Materials under Combined Repeated Stresses Superimposed with Static Stresses*, Technical Note 3495, NACA, Washington, DC.

Skertchley, S.B.J. (1879) *On the Manufacture of Gunflints*, Mem. Geol. Survey, H.M. Stationary Office, London.

Slaughter, W.S. and Fleck, N.A. (1993) Compressive fatigue of fiber composites, *Journal of Mechanics and Physics of Solids*, **41**, 1265 – 84.

Soderberg, C.R. (1939) Factor of safety & working stress, *Trans. American Society of Mechanical Engineering*, **52**, 13 – 28.

Sokolnikoff, I.S. (1956) *Mathematical Theory of Elasticity*, McGraw-Hill, New York.

Speth, J.D. (1972) Mechanical basis of percussion flaking, *American Antiquity*, **37** (1), 60.

*Standard Recommended Practices*, ASTM Designations E 466 – 468, American Society for Testing and Materials, Philadelphia.

Starkey, M.S. and Skelton, R.P. (1982) A comparison of the strain intensity and cyclic *J* approaches to crack growth, *Fatigue of Engineering Materials and Structures*, **5**, 329 – 41.

Stephenson, R. (1849) Discussion of the paper On Railway Axels by J.E. McConnel at an October 24, 1849 meeting of the Institute of Mechanical Engineers, London. Quoted from Timoshenko, S.P. (1953), p. 164.

Strong, A.B. (1989) *Fundamentals of Composites Manufacturing: Materials, Methods and Applications*, Society of Manufacturing Engineers, Dearborn, Michigan.

Sun, C.T. and Chu, G.D. (1991) Reducing free edge effect on laminate strength by edge modification, *Journal of Composite Materials*, **25**, 142 – 161.

Suresh, S. (1990) Fatigue crack growth in ceramic materials at ambient and elevated temperatures, in *Proceedings 4th International Conference on Fatigue and Fatigue Thresholds* (eds H. Kitagawa and T. Tanaka), Vol. II, MCEP, Birmingham, UK, pp. 759 – 68.

Suresh, S. (1991) *Fatigue of Materials*, Cambridge University Press, Cambridge, UK.

Suresh, S., Zamiski, G.F. and Ritchie, R.O. (1981) Oxide induced crack closure: an explanation for near threshold corrosion fatigue crack growth behavior, *Metallurgical Transactions*, **12A**, 1435 – 43.

Swain, M.H., Everett, R.A., Newman, J.C. Jr., *et al.* (1988) The growth of small cracks in 4340 Steel, in *Proceedings American Helicopter Society National Technical Specialists Meeting on Advanced Rotorcraft Structures*, American Helicopter Society, Williamsburg, Va.

Swain, M.H. (1992) Monitoring small-crack growth by the replication methods, in *Small-Crack Test Methods* (eds J.M. Larsen and J.E. Allison) ASTM STP 1149, American Society for Testing and Materials, Philadelphia, pp. 34 – 56.

Tack, A.J. and Beevers, C.J. (1990) The influence of compressive loading on fatigue crack propagation in three aerospace bearing steels, in *Proceedings of 4th International Conference on Fatigue and Fatigue Thresholds*, (eds H. Kitagawa and T. Tanaka), MCEP, Birmingham, UK, pp. 1179 – 84.

Tada, H., Paris, P.C. and Irwin, G.R. (1973) *Stress Analysis of Cracks Handbook*, Del Research Corp., Hellertown, Pa.

Takemori, M.T. (1982) Fatigue fracture of polycarbonate, *Polymer Engineering and Science*, **22**, 937 – 45.

Takemori, M.T. (1984) Polymer Fatigue, *Annual Review of Materials Science*, **14**, 171 – 204.

Talreja, R. (1986) Stiffness properties of composite laminates with matrix cracking and interior delamination, *Engineering Fracture Mechanics*, **25**, 751 – 62.

Tanaka, K. (1974) Fatigue crack propagation from a crack inclined to the cyclic tensile axis, *Engineering Fracture Mechanics*, **6**, 493 – 507.

Tanaka, K., Nakai, Y. and Yamashita, M. (1981) Fatigue growth threshold of small cracks, *International Journal of Fatigue*, **17**, 519 – 33.

Taylor, D. (1985) *A Compendium of Fatigue Thresholds and Growth Rates*, Chameleon Press, London.

Taylor, D. (1990) Fatigue in biomaterials, *in Proceedings 4th International Conference on Fatigue and Fatigue Thresholds* (eds H. Kitagawa and T. Tanaka) MCEP, Birmingham, UK, pp. 55 – 64.

Taylor, D. and Knott, J.F. (1981) Fatigue crack propagation behavior of short cracks: the effect of microstructure, *Fatigue of Engineering Materials and Structures*, 4, 147 – 55.

ten Have, A.A. (1989) European approaches in standard spectrum development, *in Development of Fatigue Loading Spectra*, (eds J.M. Potter and R.T. Watanabe) ASTM STP 1006, American Society Testing and Materials, pp, 17 – 35.

Thompson, E.R. and Lempkey, F.D. (1974) Directionally solidified eutectic superalloys, in *Composite Materials*, 4, (ed. K.G. Kreider) Academic Press, London pp, 101 – 57.

Thompson, N., Wadsworth, N.J., and Louat, N. (1956) The origin of fatigue fracture in copper, *Philosophic Magazine*, 1, 113 – 26.

Timoshenko, S.P. (1953) *History of Strength of Materials*, McGraw-Hill, New York.

Timoshenko, S.P. and Gere, J.M. (1961) *Theory of Elastic Stability*, 2nd edn, McGraw-Hill, New York.

Timoshenko, S.P. and Goodier, J.N. (1970) *Theory of Elasticity*, 3rd edn, McGraw-Hill, New York.

Tohgo, K., Otsuka, A. and Yoshida, M. (1990) Fatigue behavior of a surface crack under mixed mode loading, in *Proceedings 4th International Conference on Fatigue and Fatigue Thresholds*, (eds H. Kitagawa and T. Tanaka) MCEP, Birmingham, UK, pp. 567 – 72.

Tong, P., Greif, R. and Chen, L. (1994) Residual strength of aircraft panels with multiple site damage, *Computational Mechanics*, 13, 285 – 94.

Topper, T.H., DuQuesnay, D.L. and Pompetzki, M.A (1992) Crack closure, damage and short crack growth under variable amplitude loading, in *Theoretical Concepts and Numerical Analysis of Fatigue* (eds. A.F. Blom and C. J. Beevers) EMAS, Warley, UK, pp. 201 – 35.

Trevelyan, O.M. (1942) *Illustrated English Social History*, Vol. 3, David McKay, New York.

Trewetthey, B.R., Gillespie, J.W. Jr. and Carlsson, L.A. (1988) Mode II cyclic delamination growth, *Journal of Composite Materials*, **22**, 459 – 83.

Tschegg, E.K. (1983) Mode III and mode I fatigue crack propagation behavior under torsional loading, *Journal of Materials Science*, **18**, 1604 – 14.

Tschegg, E.K. and Stanzl, S. (1981) Fatigue crack propagation and threshold in BCC and FCC metals at 77 and 293 K, *Acta Metallurgica*, **29**, 33 – 40.

Tschegg, E.K., Stanzl, S.E., Mayer, H.R. *et al.* (1992) Crack face interactions and near-threshold fatigue crack growth, *Fatigue and Fracture of Engineering Materials and Structures*, **16**, 71 – 83.

Tvergaard, V. (1990) Micromechanical modeling of fiber debonding in a metal reinforced by short fibers, in *Inelastic Deformation of Composite Materials* (ed. G.J. Dvorak), Springer-Verlag, Berlin, pp. 99 – 114.

Tylecote, R.F. (1976) *A History of Metallurgy*, The Metals Society, London.

Van Vlack, L.H. (1985) *Elements of Materials Science*, 2nd edn, Addison-Wesley, Reading, Mass.

Vasudevan, A.K. and Suresh, S. (1982) Influence of corrosion deposits on near threshold fatigue crack growth behavior in 2XXX & 7XXX series aluminum alloys, *Metallurgical Transactions*, **13A**, 2271 – 80.

Vinson, J.R. ed. (1978) *Advanced Composite Materials – Environmental Effects*, ASTM STP 658, American Society for Testing and Materials, Philadelphia.

Wagner, H. D. (1989) Statistical concepts in the study of fracture properties of fibres and composites, in *Application of Fracture Mechanics to Composite Materials* (ed. K. Friedrich), Elsevier Science Publishers, Amsterdam, pp. 39 – 77.

Wagner, L., Gregory, J.K., Gysler, A. and Lütjering, G. (1986) Propagation behavior of short cracks in Ti-8.6 Al alloy, in *Small Fatigue Cracks* (eds R.O. Ritchie and J. Lankford), Metallurgical Society of American Institute of Mining, Mineral and Petroleum Engineers, Warrendale, Pa, pp. 117 – 28.

Walker, N. and Beevers, C.J. (1979) A fatigue crack closure mechanism in titanium, *Fatigue in Engineering Materials and Structures*, **1**, 135 – 48.

Wanhill, R.J.H. (1986) Short cracks in aerospace structures, in *The Behavior of Short Fatigue Cracks* (eds K.J. Miller and E.R. de los Rios), Mechanical Engineering Publ., London, pp. 27 – 36.

Wanhill, R.J.H. (1992) *A Preliminary Study of Fatigue Durability in Terms of Short and Long Crack Growth*, NLR TP 92437L, National Aerospace Laboratory, Amsterdam, The Netherlands.

Ward-Close, C.M. and Ritchie, R.O. (1988) On the role of crack closure mechanisms in influencing fatigue crack growth following tensile overloads in titanium alloys, in *Mechanisms of Fatigue Crack Closure* (eds J.C. Newman and W. Elber), ASTM STP 982, American Society for Testing and Materials, Philadelphia, pp. 93 – 111.

Ward-Close, C.M., Blom, A.F. and Ritchie, R.O. (1989) Mechanisms associated with transient fatigue crack growth under variable amplitude loading: an experimental and numerical study, *Engineering Fracture Mechanics*, **324**, 613 - 38.

Warren, R. and Sarin, V.K. (1989) Fracture of whisker-reinforced ceramics, in *Application of Fracture Mechanics to Composite Materials* (ed. K. Friedrich), Elsevier Science Publ., Amsterdam, pp. 571 – 614.

Waterhouse, R.B. (1972) *Fretting Corrosion*, Pergamon Press, London.

Weertman, J. (1966) Rate of growth of fatigue cracks calculated from the theory of infinitesimal dislocations distributed on a crack plane, *International Journal of Fracture*, **2**, 460 – 7.

Wei, R.P. and Landes, J. (1970) Correlation between sustained load and fatigue crack growth in high strength steels, *Materials Research and Standards*, **9**, 25-7, 44 – 6.

Wei, R.P. and Simmons, G.W. (1981) Recent progress in understanding environment assisted fatigue crack growth, *International Journal of Fracture*, **17**, 235 – 47.

Weibull, W. (1939) A statistical theory of the strength of materials, *Proceedings of Royal Swedish Academy of Engineering Sciences*, **151**, Stockholm.

Weibull, W. (1961) *Fatigue Testing and Analysis of Results*, Pergamon Press, Oxford.

Weller, T. and Singer, J. (1990) Durability of stiffened composite panels under repeated buckling, *International Solids and Structures*, **26**, 1037 – 69.

Westergaard, H.M. (1939) Bearing pressures and cracks, *Journal of Applied Mechanics*, **6**, 49 – 53.

Wheeler, O.E. (1972) Spectrum loading and crack growth, *Journal of Basic Engineering*, **94**, 181 – 7.

Whitney, J.M., Browning, C.E.and Hoogstederr, W. (1982) A double cantilever beam test for characterizing mode I delamination of composite materials, *Journal of Reinforced Plastics and Composites*, **1**, 297 – 313.

Willenborg, J., Engle, R. and Wood, H. (1971) *A Crack Growth Retardation Model Using an Effective Stress Concept*, US Air Force Flight Dynamics Laboratory – TM-71.

Williams, M.L. (1957) On the stress distribution at the base of a stationary crack, *Journal of Applied Mechanics*, **24**, 109 – 14.

Williams, M.L. (1961) The bending distribution at the base of a stationary crack, *Journal of Applied Mechanics*, **28**, 78-82.

Wright, T.M. and Hayes, W.C. (1976) The fracture mechanics of fatigue crack propagation in compact bone, *Journal of Biomedical Materials Research* **7**, 637 – 48.

Wright, T.M. and Robinson, R.P. (1982) Fatigue crack propagation in poly (methyl methacrylate) bone cements, *Journal of Materials Science*, **17**, 2463 – 8.

Yang, J.N. and Manning, S.D. (1990) Stochastic crack growth analysis methodologies for metallic structures, *Journal of Engineering Fracture Mechanics*, **37**, 1105 – 24.

Yang, J.M., Thayer, R.B., Chen, S.T. *et al.* (1991) Creep of fiber-reinforced ceramic matrix composites, in *Proceedings 8th International Conference on Composite Materials* (eds S.W. Tsai and G.S. Springer), Society for the Advancement of Materials and Process Engineering, Covina, Ca, pp. 23C1 – 11.

Yoon, K.B., Huh, Y.H. and Saxena, A. (1993) Creep-fatigue crack growth behavior during load hold period with preceding or succeeding cyclic overload, in *Proceedings 5th International Conference on Fatigue and Fatigue Thresholds* (eds J.P. Bailon and J.I. Dickson) EMAS, Warley, UK, pp. 891 – 96.

Young, A., Cartwright, D.J. and Rooke, D.P. (1988) The boundary element method for analysing repair patches on cracked finite sheets, *Journal of Royal Aeronautical Society*, **92**, 416 – 21.

Yu, M.T., Topper, T.H. and Au, P. (1984) The effects of stress ratio, compressive load and underload on the threshold behavior of a 2024-T351 aluminum alloy, in *Proceedings 2nd International Conference on Fatigue and Fatigue Thresholds*, (ed. C.J. Beevers) EMAS, Warley, UK, pp. 179 – 90.

Yu, M.T. and Topper, T.H. (1985) The effects of material strength, stress ratio, and compressive overload on the threshold behavior of a SAE 1045 steel, *Journal of Engineering Materials and Technology*, **107**, 19 – 25.

Yu, M.T., Topper, T.H., DuQuesnay, D.L. *et al.* (1986) The effect of compressive peak stress on fatigue behavior, *International Journal of Fatigue*, **8**, 9 –15.

Zaiken, E. and Ritchie, R.O. (1985) On the role of compressive overloads in influencing crack closure and the threshold condition for fatigue crack growth in 7150 aluminum alloy, *Engineering Fracture Mechanics*, **22**, 35 – 48.

# Index